高等职业教育系列教材

大数据分析技术

李俊翰　聂　强　主　编
付　雯　于　鹏　副主编
曹长勇　周文博　参　编

机械工业出版社

本书主要面向高职大数据技术专业的学生，注重大数据分析技术的应用和实践。本书每个项目主要分为两个部分。第一部分是大数据分析技术的理论知识，主要讲解了大数据分析模型、Python 数据分析工具、NumPy 和 Pandas 数据分析库、Matplotlib 数据分析可视化库、Hadoop 及其常用组件以及 scikit-learn 机器学习库的基本原理和操作。第二部分是任务实施，通过 17 个实操任务充分展现了大数据分析技术的主要功能和特点。

本书既可作为高等职业院校大数据技术、信息安全与管理、软件技术、计算机网络技术、云计算技术等专业的教材，也适合有一定 Python 编程经验并对大数据分析技术感兴趣的读者阅读。

本书配有微课视频，扫描书中二维码即可观看。另外，本书配有电子课件、源代码和习题答案，需要的教师可登录机械工业出版社教育服务网（www.cmpedu.com）免费注册，审核通过后下载，或联系编辑索取（微信：13261377872，电话：010-88379739）。

图书在版编目（CIP）数据

大数据分析技术 / 李俊翰，聂强主编. —北京：机械工业出版社，2022.8
高等职业教育系列教材
ISBN 978-7-111-71208-4

Ⅰ. ①大… Ⅱ. ①李… ②聂… Ⅲ. ①数据处理-高等职业教育-教材
Ⅳ. ①TP274

中国版本图书馆 CIP 数据核字（2022）第 125610 号

机械工业出版社（北京市百万庄大街 22 号　邮政编码 100037）
策划编辑：王海霞　　责任编辑：王海霞　李培培
责任校对：张艳霞　　责任印制：邸　敏

中煤（北京）印务有限公司印刷

2022 年 8 月第 1 版・第 1 次印刷
184mm×260mm・14 印张・345 千字
标准书号：ISBN 978-7-111-71208-4
定价：59.00 元

电话服务　　　　　　　　　　　　　网络服务
客服电话：010-88361066　　　　　　机　工　官　网：www.cmpbook.com
　　　　　010-88379833　　　　　　机　工　官　博：weibo.com/cmp1952
　　　　　010-68326294　　　　　　金　书　网：www.golden-book.com
封底无防伪标均为盗版　　　　　　　机工教育服务网：www.cmpedu.com

Preface 前言

随着大数据时代的加速发展，各种物联网智能设备产生了大量数据，如何获取、处理、分析和可视化呈现数据，将对各行各业的发展和人们的日常生活产生极大的影响。大数据分析是信息技术领域的一场革命。从越来越多的企业开始设立大数据分析岗位就可以看出，大数据分析人才已经成为各大企业竞相争夺的重要资源。为了进一步满足社会对大数据分析人才的需求，许多高校相继开设了大数据技术专业，并将大数据分析作为专业平台课程进行推广，满足不同群体在诸多领域如"1+X"证书培训、大数据技术技能竞赛、大数据分析技术专业课程，以及大数据分析技术岗位等的不同需求。

为了能够更好地满足大数据技术专业对数据分析及工具应用的技能需求，本书编写组与新华三集团深度合作，以推动"岗课赛证融通"，培养高技能人才为目的，以"项目分析→知识准备→任务实施→练习题"为主线编写。全书"理实一体"，便于实施"做中学，学中做"的教学方法。

本书为国家教学资源库"大数据分析技术"课程的配套教材，读者可以通过智慧职教网站加入在线课程的学习。

主要读者对象

- 高等职业院校大数据技术专业和人工智能专业的学生。
- 大数据分析技术初学者。
- 大数据可视化工程师。
- 大数据及数据挖掘工程师。
- 其他对 Python 或 Hadoop 框架感兴趣的人员。

主要内容

本书采用项目驱动的方式，从理论到实践，由浅入深地设计了 7 个项目，包含 17 个任务，具体内容如下。

项目 1 介绍了大数据分析技术的概念和发展、作用和影响、技术和工具以及处理流程，然后介绍大数据分析常用的 4 种分析模型，最后介绍两个大数据分析的实例：网站用户活跃程度指标综合分析和身高体重指标数据关联分析。

项目 2 首先介绍了 Python 数据分析的特点和 Python 数据分析常用库，然后介绍 Anaconda 开源个人版（发行版）及其安装步骤，接着介绍了 PyCharm 社区版及其安装步骤和运行方式，最后实现 Anaconda 开源个人版（发行版）、Jupyter Notebook 和 PyCharm 社区版的安装和运行。

项目 3 首先介绍了 NumPy 的基础理论和引用方法，然后介绍 NumPy 的 ndarry 对象、数据类型、数组和属性，以及常用函数、广播机制和方法，最后介绍使用 NumPy 实现股票数据分析和豆瓣电影数据分析。

项目 4 首先介绍了 Pandas 的基础理论和引用方法，然后介绍 Pandas 对象、基本数据类型、索引和切片、缺失值和空值处理、连接和合并数据、分组、重塑、数据透视表、时间序列、分类，以及 IO 操作，最后介绍了 3 个 Pandas 的数据分析案例，分别实现水果销售数据分析、用户消费行为数据分析和电商销售数据分析。

项目 5 首先介绍了 Matplotlib 的基础理论和引用方法，然后介绍并实现了 Matplotlib 的几种常见图形，最后使用饼图、折线图实现零售总额数据分析和双条形图实现零售总额变化情况数据分析。

项目 6 首先介绍了 Hadoop 的核心组件和工作原理，然后介绍了 Hadoop 生态圈组件，最后使用 Hadoop 及其组件 Hive，以及 Spark 实现数据分析。

项目 7 首先介绍了机器学习基本概念，然后介绍 scikit-learn 的基本用法，以及使用 scikit-learn 实现鸢尾花数据分析和波士顿房价数据分析。

编写分工

本书为新形态一体化教材，配套建设了电子课件、微课视频、源代码和习题答案等数字化学习资源。

本书由重庆电子工程职业学院李俊翰、聂强担任主编，重庆电子工程职业学院付雯以及新华三集团于鹏担任副主编，重庆电子工程职业学院大数据专业曹长勇、周文博担任参编。李俊翰负责项目 3~项目 7 的编写，聂强负责项目 1 的编写，付雯、于鹏负责项目 2 的编写，曹长勇、周文博负责部分案例的编写。

由于编者的水平有限，书中难免有一些错误或不准确的地方，恳请各位读者不吝指正。本书涉及的所有技术内容都只能用于教学，不能用于其他用途。

编　者

目录 Contents

前言

项目 1　认识大数据分析 ……………………………………………… 1

1.1 大数据分析概述 …………………… 1
1.1.1 大数据分析的概念和发展 …… 1
1.1.2 大数据分析的作用和影响 …… 2
1.2 大数据分析模型 …………………… 3
1.2.1 大数据分析的技术和工具 …… 3
1.2.2 大数据分析的流程 …………… 4
1.2.3 大数据分析的模型简介 ……… 5
【任务实施】 …………………………… 6
任务 1　网站用户活跃度指标综合分析 … 6
任务 2　身高体重指标数据关联分析 … 8
练习题 ………………………………… 11

项目 2　安装 Python 数据分析工具 ……………………………… 12

2.1 Python 数据分析基本概念 ……… 12
2.1.1 Python 数据分析的特点 …… 12
2.1.2 Python 与其他数据分析工具的比较 … 12
2.2 Python 数据分析常用库介绍 …… 13
2.2.1 NumPy 简介 ………………… 13
2.2.2 Pandas 简介 ………………… 13
2.2.3 SciPy 简介 …………………… 14
2.2.4 Matplotlib 简介 ……………… 14
2.2.5 scikit-learn 简介 ……………… 14
2.2.6 Statmodels 简介 ……………… 14
2.2.7 Seaborn 简介 ………………… 15
【任务实施】 …………………………… 15
任务 1　在 Windows 系统中安装 Anaconda … 15
任务 2　运行 Jupyter Notebook …… 19
任务 3　PyCharm 的安装和使用 …… 21
练习题 ………………………………… 26

项目 3　使用 NumPy 实现统计分析和处理 ……………………… 27

3.1 NumPy 的基本概念 ……………… 27
3.1.1 NumPy 基础理论和引用方法 … 27
3.1.2 ndarry 对象 ………………… 28
3.1.3 NumPy 数据类型 …………… 29
3.1.4 NumPy 数组属性 …………… 30
3.1.5 NumPy 切片和索引 ………… 31
3.2 NumPy 函数 ……………………… 32
3.2.1 NumPy 数学函数 …………… 32
3.2.2 NumPy 数组维度操作函数 … 36
3.2.3 NumPy 创建数组函数 ……… 39
3.2.4 NumPy 常用 IO 函数 ……… 41
3.2.5 NumPy 广播 ………………… 44
【任务实施】 …………………………… 46
任务 1　使用 Numpy 实现股票数据分析 … 46
任务 2　使用 Numpy 实现豆瓣电影数据分析 …………………………… 47
练习题 ………………………………… 51

项目 4　Pandas 数据分析和处理 ……………… 52

- 4.1　Pandas 的基本概念 …………… 52
  - 4.1.1　Pandas 基础理论和引用方法 …… 52
  - 4.1.2　Pandas 基本数据结构 ………… 53
- 4.2　Pandas 的基本用法 …………… 53
  - 4.2.1　创建 Pandas 对象 …………… 53
  - 4.2.2　查看 Pandas 基本数据 ……… 56
  - 4.2.3　Pandas 索引和切片 ………… 59
  - 4.2.4　Pandas 缺失值和空值处理 …… 62
  - 4.2.5　Pandas 连接和合并数据 …… 63
  - 4.2.6　Pandas 分组 ………………… 66
  - 4.2.7　Pandas 重塑 ………………… 67
  - 4.2.8　Pandas 数据透视表 ………… 69
  - 4.2.9　Pandas 时间序列 …………… 70
  - 4.2.10　Pandas 分类 ……………… 73
  - 4.2.11　Pandas IO 操作 …………… 73
- 【任务实施】……………………………… 74
  - 任务 1　使用 Pandas 实现水果销售数据分析 ……………………………… 74
  - 任务 2　使用 Pandas 实现用户消费行为数据分析 ………………………… 84
  - 任务 3　使用 Pandas 实现电商销售数据分析 … 94
- 练习题 …………………………………… 104

项目 5　Matplotlib 数据分析可视化库 ……………… 105

- 5.1　Matplotlib 的基本概念 ………… 105
  - 5.1.1　Matplotlib 基础理论和引用方法 … 105
  - 5.1.2　散点图 ……………………… 106
  - 5.1.3　条形图 ……………………… 108
  - 5.1.4　折线图 ……………………… 111
  - 5.1.5　饼图 ………………………… 114
  - 5.1.6　直方图 ……………………… 117
  - 5.1.7　箱形图 ……………………… 120
- 5.2　组合图 …………………………… 121
  - 5.2.1　曲线组合图 ………………… 121
  - 5.2.2　柱状、散点、折线组合图 …… 122
  - 5.2.3　直方图组合图 ……………… 123
- 【任务实施】……………………………… 124
  - 任务 1　使用饼图实现零售总额数据分析 … 124
  - 任务 2　使用折线图实现零售总额数据分析 ………………………………… 126
  - 任务 3　使用双柱状图实现零售总额变化情况数据分析 …………………… 127
- 练习题 …………………………………… 130

项目 6　基于 Hadoop 的数据分析 ……………… 131

- 6.1　掌握 Hadoop 框架和生态组件 …… 131
  - 6.1.1　Hadoop 简介 ………………… 131
  - 6.1.2　Hadoop 核心组件和工作原理 … 132
  - 6.1.3　Hadoop 安装、部署和应用 …… 136
- 6.2　Hadoop 生态组件 ……………… 147
  - 6.2.1　Hadoop 生态圈简介 ………… 147
  - 6.2.2　Hive 的安装、部署和应用 …… 148
  - 6.2.3　Spark 的安装、部署和应用 …… 158
  - 6.2.4　HBase 的安装、部署和应用 … 163
  - 6.2.5　Kafka 的安装、部署和应用 …… 168
  - 6.2.6　Flume 的安装、部署和应用 … 172
  - 6.2.7　Sqoop 的安装、部署和应用 … 178
  - 6.2.8　Zookeeper 的安装、部署和应用 … 183

【任务实施】……………………………186	任务 2 使用 Hadoop 及其组件 Spark 实现
任务 1 使用 Hadoop 及其组件 Hive 实现数据	数据分析………………………189
分析…………………………186	练习题………………………………191

项目 7 基于 scikit-learn 机器学习库的数据分析……192

7.1 掌握机器学习基本概念…………192	7.2.2 scikit-learn 的基本用法……………205
7.1.1 机器学习简介………………………192	【任务实施】……………………………206
7.1.2 机器学习基本流程…………………193	任务 1 使用 scikit-learn 实现鸢尾花数据
7.1.3 机器学习开发流程…………………194	分析…………………………206
7.1.4 机器学习算法分类…………………194	任务 2 使用 scikit-learn 实现波士顿房价
7.2 掌握 scikit-learn 的基本用法……204	数据分析……………………210
7.2.1 scikit-learn 的安装和引用方法……204	练习题………………………………215

参考文献……………………………………………………216

微课视频清单

名称	二维码	名称	图形
1-1 用户活跃度分析		4-3 使用 pandas 实现电商销售数据分析	
1-2 身高体重关联分析		5-1 使用饼图实现零售总额数据分析	
2-1 在 Windows 系统中安装 Anaconda		5-2 使用折线图实现零售总额数据分析	
2-2 PyCharm 的安装和使用		5-3 使用双柱状图实现零售总额变化情况数据分析	
3-1 使用 Numpy 实现股票数据分析		6-1 使用 Hadoop 及其组件 HIVE 实现数据分析	
3-2 使用 Numpy 实现豆瓣电影数据分析		6-2 使用 Hadoop 及其组件 Spark 实现数据分析	
4-1 使用 pandas 实现水果销售数据分析		7-1 鸢尾花数据分析	
4-2 使用 pandas 实现用户消费行为数据分析		7-2 波士顿房价预测	

项目 1　认识大数据分析

【项目分析】

本项目旨在帮助读者认识大数据分析的基本框架，为后续大数据分析的学习做好铺垫。具体内容如下。

1）大数据分析、大数据的基本概念。

2）大数据分析的基础：数据挖掘与分析、数据可视化分析、预测分析、语义分析以及数据分析和质量管理。

3）从帮助提升工作效率、提高数据和业务的可理解性、可读性以及精准性方面讲解大数据分析的作用和影响。

4）大数据分析的技术和工具。

5）大数据分析的流程、模型及其案例。

【知识准备】

1.1　大数据分析概述

1.1.1　大数据分析的概念和发展

随着物联网、社交媒体和电子商务等业务爆发式地增长，数据"洪流"呈指数级地从四面八方不断涌来。大数据因其数据量大（Volume）、速度快（Velocity）、类型多（Variety）、价值高（Value）、真实性（Veracity）的特点，已经成为各行各业竞相追逐的重要领域，大数据分析也应运而生。

大数据分析是指对海量数据进行的数据分析。大数据是大数据分析的基石，海量的数据再多，如果没有大数据分析，也只是存放在存储设备中的无用数据而已，难以对数据的深层价值进行深入挖掘。

大数据分析的基础包含以下 5 个方面，如图 1-1 所示。

(1) 数据挖掘与分析

从广义的角度出发，数据挖掘算法是大数据分析的重要核心内容。只有用特定的数据挖掘算法处理不同业务场景所产生的数据类型和结构，才能获得高质量的价值数据。同时，随着数据挖掘算法的不断优化，使其能够处理更多、更大和更复杂的数据内容。

图 1-1　大数据分析的基础

(2) 数据可视化分析

数据可视化分析因其直观、易读、易理解，不仅被大数据分析科学家使用，也大量应用在普通客户的业务需求之中。数据可视化分析能够非常高效地将晦涩难懂、抽象的数据，以清晰、直接的各种图形和表格形式非常简单地描述出数据背后所蕴含的丰富故事。

(3) 预测分析

大数据预测分析是大数据分析的重要应用，通过应用数据挖掘算法找到特定业务领域的大数据特点，并建立符合行业特征的数据模型，实现对未来业务数据进行有效的预测，帮助优化不同行业、企业的经营和决策。

(4) 语义分析

大数据语义分析用于对网络数据的分析和挖掘，通过对用户不同行为产生的不同数据，例如评论关键词、搜索关键词、产品关键词等特定行业的上下文语义，有针对性地分析和判断用户需求和行为模式，为企业和用户提供更好的服务和体验。

(5) 数据分析和质量管理

大数据不仅仅只有海量的数据，更需要高质量的数据和数据管理作为重要支撑。高质量的数据能够更加精准地针对特定业务数据提供更有价值的数据分析结果。

以上是大数据分析的 5 个基础方面，随着大数据分析技术的不断发展，在未来还会有更多、更好、更专业的大数据分析技术。

1.1.2 大数据分析的作用和影响

大数据分析技术已经融入了社会发展的方方面面，有着举足轻重的作用，实现了巨大的价值。随着劳动力市场对数据分析人才的需求与日俱增，越来越多的企业专门成立了数据分析部门。大数据分析技术在政治、经济、社会等各领域发挥着越来越重要的作用。具体作用如图 1-2 所示。

(1) 进一步提升工作效率

面对需要处理的海量数据，不仅需要花费大量的人力、物力和财力对其进行有效存储、管理和维护，还要对其实施有效的分析处理才能发现变量和常量等数据之间隐藏的内在关联。因此，数据分析能够通过正确的呈现方式，将数据之间千丝万缕的关系和规律进行简单的描述，从而提升工作效率。

(2) 让业务和数据变得更加易于理解、可读

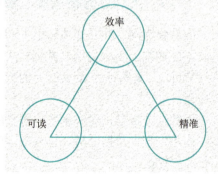

图 1-2 大数据分析的作用

传统的数据库管理系统往往不能非常直观地呈现出数据整体和局部的状态和联系。不仅对于专业数据科学家，还包括普通用户，要及时、高效地理解和掌握某个领域或业务当中海量的结构化数据表，甚至是半结构化和非结构化的数据，都是十分困难的事情。通过应用大数据分析技术能够让复杂的数据变得可读、可理解，有利于数据工作者和客户对数据进行更优化的归纳和总结，并使得特定的业务数据变得更加有逻辑，条理更加清晰。

(3) 业务开展更加精准

使用大数据分析技术的目的就是在海量数据之中发现知识，用科学的统计方法帮助人们精准定位业务过程中可能存在的问题和机会，从而避免业务实施过程中的失误，使得业务实施的

路线和方向更加明确和精准。

大数据分析技术及其工具已经应用到了各行各业，以数据驱动业务的理念已经成为行业主流的工作方式，大数据分析技术让业务开展更加高效和精准。

1.2 大数据分析模型

1.2.1 大数据分析的技术和工具

1. Python

Python 拥有非常成熟的技术和资源社区，能够在数据分析和处理、人工智能以及数据可视化等方面提供强大的技术支持。大数据业务流程的数据采集、数据存储、数据分析、数据处理，以及数据可视化都是应用 Python 及其类库实现的。比较著名的类库和工具有 NumPy、Pandas、Matplotlib、scikit-learn 等。

（1）Numpy

Numpy 是 Python 中的一个矩阵计算包，提供了非常强大的数组 ndarray 及其相应的处理函数 ufunc，使其能够在高效地应用内存的前提下，非常快速地创建 N 维数组，并提供矢量化数学运算，可在不编写特定循环的情况下对整个数组进行标准运算操作。NumPy 的出现不仅极大地弥补了 Python 在操作列表数据类型进行数值运算过程中比较耗费内存和 CPU 计算资源的问题，而且很好地补充了 Python 的 array 模块不能支持多维数组及其运算的不足。

（2）Pandas

Pandas 是建立在 NumPy 之上的一个 Python 数据分析包，拥有强大的数据分析功能，包含 Series、DataFrame 等高级数据结构和工具。Pandas 纳入了大量的库和一些标准的数据模型，提供了高效操作大型数据集所需要的工具。Pandas 提供了大量快速便捷的函数和方法。其中，Pandas 拥有的坐标轴特点能够支持数据自动对齐，高效提升不同数据源的数据探索和检索，防止数据处理过程中可能出现的问题。

（3）Matplotlib

Matplotlib 是基于 NumPy 的一套 Python 包，拥有非常丰富的可视化图形库和工具，通过强大的接口能够绘制满足不同数据类型和业务需求的专业图形，并且能支持多种操作系统的不同 GUI 后端，输出的图形格式可以是 PDF、SVG、JPG、PNG、BMP、GIF 等。

（4）scikit-learn

scikit-learn（简称 sklearn）是一个基于 Python 语言的机器学习开源框架，包含了大量优质的数据集，在学习机器学习的过程中，可以通过使用这些数据集实现出不同的模型。

2. Hadoop 及其主要生态圈

Hadoop 是大数据技术重要代表之一。它来源于 Apache 基金会以 Java 编写的开源分布式框架项目。其核心组件是 HDFS、YARN 和 MapReduce，其他组件为 HBase、Hive、ZooKeeper 和 Sqoop 等。这些组件共同提供了一套完整服务或更高级的服务。Hadoop 可以将大规模海量数据进行分布式并行处理。Hadoop 具有高度容错性、可扩展性、高可靠性和稳定性，这些优势让

Hadoop 成为当下最流行的大数据应用平台之一。

1.2.2 大数据分析的流程

科学合理的大数据分析流程会极大地影响数据分析的质量和效率。大数据分析的流程具体分为 5 个步骤，如图 1-3 所示。

图 1-3　大数据分析的流程

（1）识别问题

在开展大数据分析之前，首先应该明确具体任务，以及需要解决的问题。只有以解决问题为导向，才能更加清晰、有效地围绕问题开展数据分析工作。

（2）数据可行性

数据可行性是指根据识别的具体问题，判断所获得的数据是否具有可靠性、可用性和过度拟合的问题。数据可行性需要把握以下 3 个细节。

1）明确数据的体量和专业领域。

2）明确抽象概念和具体指标数据的映射关系。

3）明确具体业务中的代表性数据。

（3）数据准备

数据准备需要具体落实每个数据的具体作用和所代表的具体内涵，可以说大数据分析的大部分时间都是花费在对数据的准备过程中。数据准备具体分数据采集和数据预处理。

1）数据采集。在这个数据爆炸的时代，不论是提供底层基础架构的云计算，还是实现各种人工智能应用，都离不开其核心的源泉——数据。由于物联网海量的穿戴设备、网络多媒体平台以及电子商务平台中的数据太多、太宽泛，人们需要通过特殊的技术和方法实现在海量的数据中搜集到真正有价值的数据，从而为下一步大数据分析业务提供数据支撑。

2）数据预处理。数据预处理主要是指在对数据开展具体分析和挖掘之前对数据进行的一些处理。数据预处理有以下 4 个任务。

1）数据清洗。根据具体业务规则制定针对性的数据清洗规则，包括检测和去除数据集中的噪声数据和无关数据，处理遗漏数据，去除空白数据域或知识背景下的空值。

2）数据集成。根据具体业务需要，将不同结构和类型的数据，如不同数据库和不同格式的普通文件有机地结合在一起，使之能够为特定业务领域提供高质量的数据共享服务。例如，数据仓库模式就是一种数据集成方式。它是面向主题的、集成的、相对稳定的数据集合。

3）数据变换。根据具体大数据分析框架或技术的要求，结合具体业务的描述，将数据进行特定的转换，使之更符合大数据分析框架的特点和业务的需求。

4）数据规约。根据业务需求对原始数据进行"量体裁衣"，实现数据既能够很好地保持完整性，又能够从数据维度、数量和体积方面得到有效的缩减。

（4）数据模型

数据模型是用经过设计之后的数据对现实世界特征进行描述和呈现。数据模型能够很好地将现实世界中需要处理的问题通过计算机能够识别的方式进行高效的处理。数据模型的建立需要结合业务模型、数据分析模型，以及专家的经验判断。另外，还要考虑当前的运算能力是否满足数据模型的运算需求。

（5）分析结果

大数据分析的结果作为重要结论分为定性和定量的评估结果，分析结果必须能够严谨、可靠且符合业务决策需求。

1.2.3 大数据分析的模型简介

建立数据分析模型是进行数据分析之前的重要内容，优秀的数据分析模型能够精准映射业务数据。数据分析模型多种多样，主要可以分为对比分析、分类分析、关联分析和综合分析 4 种方式。其中，对比分析、分类分析和关联分析以定性分析为主，综合分析以定性和定量分析相结合为主。

（1）对比分析模型

对比分析模型是指在特定业务背景知识中将多个数据进行比较，从而发现和揭示事物的变化发展规律。对比分析能够相对简单地比较数据之间的联系，分析操作步骤较少，能够非常直观地呈现不同数据的异同，能够精准表示数据之间度量值的差距。对比标准是对比分析的主要抓手，将对比对象的指标与标准进行对比，就能得出结果了。目前常用标准是时间标准、空间标准、特定标准。例如，某年 10 月某商品的销售数量和上一年 10 月的销售数量进行对比就是基于时间标准的对比分析；不同城市之间对于共享单车的接受程度就是基于空间标准的对比分析；应用专家经验判断的数据标准与当前领域数据的对比就是基于特定标准的对比分析。

（2）分类分析模型

分类分析模型是指通过给予不同类型的数据不同的成员资格，从而将数据分成不同类别的群体，其目的是将未知类别的数据更好地向某一个类进行归纳，并按其接近归纳的程度细化分类质量。例如，通过分类分析模型区别垃圾邮件和非垃圾邮件；通过某人的某些数据指标（血糖、血压、心率等）判断其是否患有糖尿病。

（3）关联分析模型

关联分析模型是指通过分析变量之间的关系发现数据之间隐含的联系，包括明确自变量和因变量的因果变化规律或者变量之间相关性、方向性和紧密程度等。作为数据挖掘的重要技术，通过分析顾客的消费数据，发现其内在偏好规则，可为商家提供有价值的销售策略。关联分析的应用案例非常多，例如，比较著名的啤酒与尿布案例，人们发现在超市里面购买尿布的男性顾客也会同时购买啤酒，因此把啤酒和尿布放在同一个货架上进行销售。

（4）综合分析模型

综合分析模型是将多个指标综合应用在复杂数据环境中，通过分析研究对象各主要部分及其特征，并以整合宏观知识结构和突出局部重点知识的思维方式进行定性或定量分析判断，将多个指标数据整合为针对某一个综合评价的指标，从而揭示和发现复杂业务数据中各业务现象之间存在的一般或总体关系。例如，分析和评价人民幸福程度、学生综合素质和某行业发展前景等。

大数据分析模型经常应用在企业的经营和管理过程中，为企业的发展发挥着重要作用。

【任务实施】

任务 1　网站用户活跃度指标综合分析

这里将使用基于 Python 的 Pandas 数据分析库完成本任务。假设某网站的用户活跃度主要由用户登录网站次数、停留时间（秒）和点击量，3 个指标进行综合分析后得到一个综合指标：用户活跃度。上述评价指标数据样例如图 1-4 所示。

	A	B	C	D	E
1	ID	登录网站次数	停留时间（秒）	点击量	用户活跃度
2	1	16	103.7	5	
3	2	13	52.6	2	
4	3	19	90.7	9	
5	4	38	546.5	17	
6	5	27	290.1	22	
7	6	64	1063.2	54	
8	7	41	813.5	61	

图 1-4　评价指标数据样例

（1）数据标准化基本概念

综合分析模型中存在多种评价指标，由于不同评价指标的性质不同，每个评价指标有着自己的量纲和数量级。如果评价指标之间存在较大差值，较大的指标数据会对综合分析结果产生较大影响，较小的指标数据产生的影响则会相对较小。因此，数据标准化的作用就是将数据按照比例进行压缩，通过将数据无量纲化为纯数据，使之控制在某个较小的区间，有效避免了不同数据单位的影响，最后加入权重因素进行比较。数据标准化常用归一化处理方法，即把数据映射到[0,1]区间。例如，min-max 标准化（min-max normalization），log 函数转换和 z-score 标准化（zero-mean normalization）等。本任务中使用 min-max 标准化对目标数据进行归一化如式（1-1）所示。

$$y_i = \frac{x_i - \min\limits_{1\leqslant j\leqslant n}\{x_j\}}{\max\limits_{1\leqslant j\leqslant n}\{x_j\} - \min\limits_{1\leqslant j\leqslant n}\{x_j\}} \tag{1-1}$$

式中，min 为该组数据的最小值；max 为该组数据的最大值；y_i 为标准化后的新值。

（2）数据读取和观察

使用 Python 的 Pandas 数据分析库读取该数据并观察前 5 行，如图 1-5 所示。

（3）计算各字段的最大值和最小值

使用 Pandas 数据分析库的 max()和 min()函数分别计算各字段的最大值和最小值，如图 1-6 所示。

```
# 导入Pandas库并修改别名为pd
import pandas as pd
```

```
#默认读取这个Excel的第一个表单
df=pd.read_excel('用户活跃度数据.xlsx')
#默认读取前5行的数据
data=df.head()
```

```
# 呈现前5行数据
data
```

	ID	登录网站次数	停留时间（秒）	点击量
0	1	16	103.7	5
1	2	13	52.6	2
2	3	19	90.7	9
3	4	38	546.5	17
4	5	27	290.1	22

图 1-5　数据读取和观察

```
# 通过max函数得到data["登录网站次数"]最大值
df["登录网站次数"].max()
```

64

```
# 通过min函数得到data["登录网站次数"]最小值
df["登录网站次数"].min()
```

13

```
# 通过max函数得到data["停留时间(秒)"]最大值
df["停留时间(秒)"].max()
```

1063.2

```
# 通过min函数得到data["停留时间(秒)"]最小值
df["停留时间(秒)"].min()
```

52.6

```
# 通过max函数得到data["点击量"]最大值
df["点击量"].max()
```

61

```
# 通过min函数得到data["点击量"]最小值
df["点击量"].min()
```

2

图 1-6　计算各字段的最大值和最小值

（4）计算各字段标准化值

使用 Python 对用户登录网站次数、停留时间（秒）和点击量 3 个指标进行 min-max 标准化，结果如图 1-7 所示。

```
# 登录次数标准化
log_times_norm = (df["登录网站次数"]-df["登录网站次数"].min())/(df["登录网站次数"].max()-df["登录网站次数"].min())
log_times_norm
```

```
0    0.058824
1    0.000000
2    0.117647
3    0.490196
4    0.274510
5    1.000000
6    0.549020
Name: 登录网站次数, dtype: float64
```

```
# 停留时间(秒)标准化
stay_time_norm = (df["停留时间(秒)"]-df["停留时间(秒)"].min())/(df["停留时间(秒)"].max()-df["停留时间(秒)"].min())
stay_time_norm
```

```
0    0.050564
1    0.000000
2    0.037700
3    0.488720
4    0.235009
5    1.000000
6    0.752919
Name: 停留时间(秒), dtype: float64
```

```
# 点击量标准化
click_times_norm = (df["点击量"]-df["点击量"].min())/(df["点击量"].max()-df["点击量"].min())
click_times_norm
```

```
0    0.050847
1    0.000000
2    0.118644
3    0.254237
4    0.338983
5    0.881356
6    1.000000
Name: 点击量, dtype: float64
```

图 1-7　标准化结果

（5）计算加权后的综合值

对指标数据赋予特定的权重值的方法有很多，例如，可以依据历史经验，将用户登录网站次数、停留时间（秒）和点击量 3 个指标分别赋予 30%、40%和 30%的权重，得出综合评价值如图 1-8 所示。

```
In [27]: # 根据权重计算用户活跃度
         user_activity = log_times_norm*0.3 + stay_time_norm*0.4 + click_times_norm*0.3
         user_activity

Out[27]: 0    0.053127
         1    0.000000
         2    0.085967
         3    0.418818
         4    0.278051
         5    0.964407
         6    0.765874
         dtype: float64
```

图 1-8　综合评价值

（6）Series 类型转换为 DataFrame 类型

为了能够给上一步计算出的用户活跃度进行排序并修改列名（这里可以看见当前数据的列名为“0”），需要将 Series 类型的 user_activity 转换为 DataFrame 类型，如图 1-9 所示。

```
In [21]: # 将Series类型的user_activity转换为DataFrame类型
         DF = pd.DataFrame(user_activity)
         DF

Out[21]:
```

	0
0	0.053127
1	0.000000
2	0.085967
3	0.418818
4	0.278051
5	0.964407
6	0.765874

图 1-9　Series 类型转换为 DataFrame 类型

（7）修改字段名字

使用 DataFrame 类型的属性 columns 将当前列的名字修改为"用户活跃度"，如图 1-10 所示。

图 1-10　修改字段名字

（8）结果分析

通过对用户活跃度进行降序操作，可以得到排名前三的用户 ID，如图 1-11 所示。

图 1-11　活跃度排名前三的用户 ID

任务 2　身高体重指标数据关联分析

这里将使用基于 Python 的 Pandas 数据分析库完成本任务。假设某群体身高体重信息包含身高和体重两个指标，观察并分析身高和体重之间的相关性，如图 1-12 所示。

（1）相关性分析基本概念

相关性分析是指对两个或多个具备相关性的变量元素进行分析，从而衡量两个变量因素的相关密切程度。相关性的元素之间需要存在一定的联系或者概率才可以进行相关性分析。常见的相关性分析方法包括相关系数、信息增益和卡方检验等。其中，相关系数中比较具有代表性的是皮尔逊相关系数 r，定义为两个变量之间的协方差和标准差的比值，如式（1-2）所示。

图 1-12　身高和体重数据

$$r = \frac{\sum_{i=1}^{n}[(X_i - \bar{X})(Y_i - \bar{Y})]}{\sqrt{\sum_{i=1}^{n}(X_i - \bar{X})^2} \cdot \sqrt{\sum_{i=1}^{n}(Y_i - \bar{Y})^2}} \qquad (1\text{-}2)$$

式中，X_i 表示 X 的第 i 个样本数据；$\bar{X}$ 表示 i 个 X 样本的均值；Y_i 表示 Y 的第 i 个样本数据；$\bar{Y}$ 表示 i 个 Y 样本的均值。

协方差表示的是两个变量总体的误差。如果两个变量的变化趋势一致，也就是说如果其中一个大于自身的期望值，另外一个也大于自身的期望值，那么两个变量之间的协方差就是正值。如果两个变量的变化趋势相反，即其中一个大于自身的期望值，另外一个却小于自身的期望值，那么两个变量之间的协方差就是负值。

皮尔逊相关系数的变化范围为-1～1。系数值为 1 表示两个变量值之间正相关，即一个变量的值越大，另一个变量的值也会越大；系数值为-1 表示两个变量之间负相关，即一个变量的值越大，另一个变量的值反而越小；系数值为 0 表示两个变量之间没有关联。皮尔逊相关系数的优点在于算法简单，易于理解；缺点在于不能直接用于离散类型数据。

（2）数据抽取和观察

使用 Python 的 Pandas 数据分析库读取该数据并观察前 5 行，如图 1-13 所示。

```
In  [2]: #使用Pandas的read_csv（）函数读取身高体重数据相关分析.txt，该数据的分隔符是'\t'
         df=pd.read_csv('身高体重数据相关分析.txt',sep = '\t')
         # 观察源数据前5行
         data = df
         data.head()
Out[2]:
```

	姓名	身高（cm）	体重(kg)
0	王强	190	85
1	杨倩	170	70
2	张丽	179	76
3	谭鸣	188	83
4	秦军	175	73

图 1-13　数据读取和观察

（3）获取连续字段

通过自定义列表：cont_col = ["身高(cm)", "体重(kg)"]获取连续字段数据，如图 1-14 所示。

（4）计算连续字段的最大值和最小值

使用 Pandas 数据分析库的 max()和 min()函数分别计算连续字段的最大值和最小值，如图 1-15 所示。

```
In  [3]: #连续字段
         cont_col=["身高（cm）","体重(kg)"]
         #输出连续型数据
         data[cont_col]
Out[3]:
```

	身高（cm）	体重(kg)
0	190	85
1	170	70
2	179	76
3	188	83
4	175	73
5	181	79
6	177	81
7	192	89
8	167	69

图 1-14　获取连续字段数据

```
In  [4]: # 通过max函数得到data["理论"]最大值
         df["身高（cm）"].max()
Out[4]: 192

In  [5]: # 通过min函数得到data["理论"]最小值
         df["身高（cm）"].min()
Out[5]: 167

In  [6]: # 通过max函数得到data["实践"]最大值
         df["体重(kg)"].max()
Out[6]: 89

In  [7]: # 通过min函数得到data["实践"]最小值
         df["体重(kg)"].min()
Out[7]: 69
```

图 1-15　连续字段的最大值和最小值

（5）计算“身高”字段标准化

通过最大值和最小值计算“身高”字段标准化，如图 1-16 所示。

```
In [9]: # "身高(cm)"字段标准化
        height_norm = (df["身高(cm)"]-df["身高(cm)"].min())/(df["身高(cm)"].max()-df["身高(cm)"].min())
        height_norm
Out[9]: 0    0.92
        1    0.12
        2    0.48
        3    0.84
        4    0.32
        5    0.56
        6    0.40
        7    1.00
        8    0.00
        Name: 身高(cm), dtype: float64
```

图1-16 计算"身高"字段标准化

（6）计算"体重"字段标准化

通过最大值和最小值计算"体重"字段标准化，如图1-17所示。

```
In [42]: # "体重(kg)"字段标准化
         weight_norm = (df["体重(kg)"]-df["体重(kg)"].min())/(df["体重(kg)"].max()-df["体重(kg)"].min())
         weight_norm
Out[42]: 0    0.80
         1    0.05
         2    0.35
         3    0.70
         4    0.20
         5    0.50
         6    0.60
         7    1.00
         8    0.00
         Name: 体重(kg), dtype: float64
```

图1-17 计算"体重"字段标准化

（7）合并连续字段

使用Pandas的concat()函数对"身高"和"体重"字段进行合并，如图1-18所示。

（8）字段相关性系数

使用corr()函数计算"身高"和"体重"字段相关性系数，如图1-19所示。

```
In [34]: # 合并连续数据标准差
         df2 = pd.concat([height_norm, weight_norm], axis=1)
         df2
Out[34]:
      身高(cm)  体重(kg)
  0     0.92    0.80
  1     0.12    0.05
  2     0.48    0.35
  3     0.84    0.70
  4     0.32    0.20
  5     0.56    0.50
  6     0.40    0.60
  7     1.00    1.00
  8     0.00    0.00
```

```
In [36]: # 使用corr()函数计算身高和体重相关性系数
         df2.corr()
Out[36]:
           身高(cm)    体重(kg)
  身高(cm)  1.000000  0.952236
  体重(kg)  0.952236  1.000000
```

图1-18 合并连续数据　　　　　　图1-19 计算"身高"和"体重"字段相关性系数

（9）可视化库导入

导入可视化库Matplotlib和Seaborn，并指定字符集，用于显示中文标签，如图1-20所示。

```
In [48]:  # 导入可视化库Matplotlib和Seaborn
          from matplotlib import pyplot as plt
          import seaborn as sns
          # 导入指定字符集，用于显示中文标签
          plt.rcParams['font.sans-serif']=['SimHei']
          plt.rcParams['axes.unicode_minus']=False
```

图 1-20　导入可视化库并指定字符集

（10）可视化呈现

应用折线图呈现"身高"和"体重"变量之间的相关性，如图 1-21 所示。

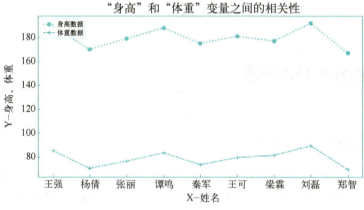

图 1-21　应用折线图呈现"身高"和"体重"变量之间的相关性

（11）结果分析

根据图 1-23，可以看见身高越高，其对应的体重数据就越大。因此，身高和体重数据之间具有极强的正相关性。

练习题

（1）大数据分析的基础包含哪 5 个方面？
（2）大数据分析有什么作用？
（3）大数据分析有哪些技术？
（4）大数据分析有哪些流程？
（5）大数据分析有哪些主要模型？

项目 2　安装 Python 数据分析工具

【项目分析】

本项目旨在理解和掌握基于 Python 编程语言的数据分析工具的特点、安装步骤和应用。具体内容如下。

1）Python 数据分析的特点，包括与其他数据分析工具之间的区别。
2）Python 数据分析常用库。
3）Anaconda 发行版及其 Jupyter Notebook 的特点、安装步骤和应用。
4）PyCharm 社区版特点、安装步骤和应用。

【知识准备】

2.1　Python 数据分析基本概念

2.1.1　Python 数据分析的特点

Python 是一种开源、灵活、简单易学、功能强大的编程语言。Python 拥有非常丰富和完善的数据分析和处理集合，可用于数据分析，目前第三方 Python 库超过 125000 个。因此，Python 在数据分析和交互、探索性计算以及数据可视化等方面都显得比较活跃，是数据分析的一大利器。这使得 Python 可以通过自身的一套技术框架解决所有的业务服务问题，充分体现了 Python 有利于各个业务之间的融合，Python 能够有效提高数据分析的效率。Python 拥有 Numpy、Matplotlib、scikit-learn、Pandas 等强大的数据分析和处理工具，这些工具都为大数据分析、处理和快速洞察提供了支持，这对企业时刻保持竞争力有很大价值，尤其是 Pandas，其在数据分析领域有着重要地位。

2.1.2　Python 与其他数据分析工具的比较

1）与 Excel 相比，Python 通过调用强大的数据分析和处理模块，灵活处理更大数据集的报表数据，并能够进一步自动地实现数据分析和建立更加复杂的机器学习模型。

2）与 R 语言过于分散和相对杂乱的机器学习库相比，Python 有着更为集中和高效的机器学习框架 scikit-learn。这让 Python 更容易被理解和掌握。因此，Python 的机器学习和数据统计分析用户社区和群体在近几年不断攀升。

3）与 SPSS 相比，Python 能够处理更为庞大和复杂的数据结构，以及适应更为复杂的数据分析业务场景。SPSS 是一款优秀的统计软件，主要应用于科学实验方面的数据分析场景。

Python 在数据科学及其一整套技术框架上面优势十分明显，包括数据采集、数据存储和管理、数据分析和处理、数据可视化、机器学习、人工智能、APP 开发和运维等整套解决方案。

2.2 Python 数据分析常用库介绍

基于 Python 的数据分析之所以强大,得益于其背后丰富的第三方库,开箱即用,方便快捷,如图 2-1 所示。

图 2-1　Python 数据分析常用库

2.2.1 NumPy 简介

NumPy 对 Python 最大的支持在于其很好地弥补了 Python 在数据组处理方面的不足,提供了强大的数值编程工具,主要用于数字处理,具有如下特点。

1)拥有矩阵数据类型、矢量处理,以及精密的运算库。

2)包括基本线性代数函数、傅里叶变换、高级随机数功能和集成工具等强大的数学科学计算工具。

3)能够高效地创建 N 维阵列,并能够通过其丰富的函数对 N 维阵列进行处理。

4)支持比 Python 内置的类型更多的数据类型。

5)底层使用 C 语言编写,数组中直接存储对象,而不是存储对象指针,所以其运算效率远高于纯 Python 代码。

2.2.2 Pandas 简介

Pandas 作为 Python 的一个开放源码程序库,不仅仅拥有许多高级的数据分析功能,而且还具备性能强大的数据分析和探索工具,具有如下特点。

1)带有高级的数据结构和精巧的工具,能够高效快速地处理数据。

2)构建在 NumPy 之上,强化和丰富了 NumPy 的使用方式。

3)能够读取各种格式的文件,包括 TXT、CSV、JSON、SQL、Excel 等。

4)能够实现对数据的自动遍历、行列查询、拆分整合、数据预处理以及各种运算操作。

5)拥有广泛的应用领域,包括科研、教育、统计、金融等数据分析领域。

2.2.3 SciPy 简介

SciPy 是一个应用于数学、科学和工程领域的开放源码的软件。它依赖于 Python 的 NumPy，拥有强大的对象和函数能够处理数据矩阵，具有如下特点。

1）高级数学计算模块包括：离散傅里叶变换、线性代数、稀疏矩阵、最优化、积分、插值、拟合、信号处理与图像处理、常微分求解方程等。

2）具有部分图形功能，能够像 MATLAB 一样绘制用于科学和工程计算的数据图形。

3）为用户提供用于操作和可视化数据的高级命令和类，为交互式 Python 会话增加了显著的功能。

4）基于 Python，从并行编程到 Web 和数据库子例程与类，Python 程序员都可以使用，可用于开发复杂的程序和专门的应用程序。

2.2.4 Matplotlib 简介

Matplotlib 是一个 Python 2D 绘图库，用于在 Python 中创建静态、动画和交互式可视化，具有如下特点。

1）可与 NumPy 一起使用，主要用于绘制二维图和部分三维图像，包括直方图、折线图和散点图等。

2）以多种硬拷贝格式和跨平台的交互式环境生成出版物质量的图形。

3）Matplotlib 可用于 Python 脚本、Python 和 IPython Shell、Jupyter Notebook、Web 应用程序服务器和四个图形用户界面。

2.2.5 scikit-learn 简介

scikit-learn 是基于 Python 编程语言的免费软件机器学习库。它依赖于 NumPy、SciPy、Matplotlib。scikit-learn 提供了一整套丰富且完善的机器学习流程和工具，具有如下特点。

1）包括数据预处理、分类、回归、聚类、预测和模型分析，包括支持向量机、随机森林、梯度提升、k 均值和 DBSCAN，并且旨在与 Python 数值科学库 NumPy 和 SciPy 联合使用。

2）与 Python 的其他库具有很好的集成性，与 Matplotlib 和 Plotly 集成可用于绘图，与 NumPy 集成可用于数组矢量化，与 Pandas 集成可用于数据分析处理等。

3）广泛使用 NumPy 进行高性能的线性代数和数组运算。

4）自带大量的数据源，方便初学者学习。

2.2.6 Statmodels 简介

Statmodels 是一个基于 Python 的模块，主要用于数据的统计分析和建模，为不同的数据类型提供了广泛统计、统计测试、绘图功能和结果统计的列表，具有如下特点。

1）Statmodels 可以与 Pandas 交互使用实现数据挖掘组合。

2）支持使用 R 风格的公式和 Pandas 的 DataFrames 数据类型。

3）与 scikit-learn 更专注于预测相比，Statsmodels 更专注于统计推理。

4）包括线性回归、广义线性模型、稳健的线性模型、线性混合效应模型等。

2.2.7 Seaborn 简介

Seaborn 是构建在 Matplotlib 之上的 Python 数据可视化库，具有如下特点。

1）通过提供基于 Matplotlib 核心库的 API 实现更为高级的统计图形绘图功能。

2）Seaborn 和 Matplotlib 相互补充能够实现更多更有特色的图形。

3）Seaborn 能够接受基于 NumPy 与 Pandas 的数据结构，并与 SciPy、Statsmodels 等统计模式高度兼容。

4）对数据矩阵进行可视化，并使用聚类算法进行分析。

5）基于时间序列的绘制和统计功能。

【任务实施】

任务 1 在 Windows 系统中安装 Anaconda

（1）Anaconda 开源个人版（发行版）简介

Anaconda 开源个人版（发行版）在全球拥有超过 2000 万名用户，是在单台机器上执行 Python 数据科学和机器学习的最简单方法。该工具包是为独立学习者开发的，可以帮助用户处理成千上万个开源软件包和库。Anaconda 包含了 Conda、Python 在内的超过 180 个科学包及其依赖项。

（2）Anaconda 特点

Anaconda 具有如下特点。

1）免费开源，强大社区支持。

2）安装简单。

3）高性能使用 Python。

4）支持多操作系统环境，多 Python 版本的切换。

（3）在 Windows 系统中安装 Anaconda

1）进入 Anaconda 官方下载地址https://www.anaconda.com/products/individual，如图 2-2 所示。

图 2-2 进入 Anaconda 官方下载地址

2）单击"Download"按钮之后进入 Anaconda 安装包选择界面，根据项目环境选择正确的操作系统环境，以及基于 Python 3.8 版本及以上的 Anaconda 安装包。这里选择基于 Windows

64 位操作系统的 Anaconda 安装包：64-Bit Graphical Installer (457 MB)，如图 2-3 所示。

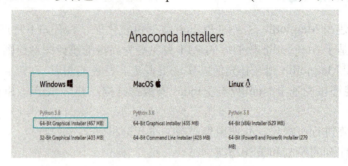

图 2-3 选择 Anaconda 安装环境

3）下载 Anaconda 安装包之后，双击该安装包进入安装欢迎界面，单击"Next"按钮，如图 2-4 所示。

4）单击"I Agree"按钮，同意 Anaconda 个人版本的最终用户许可协议，如图 2-5 所示。

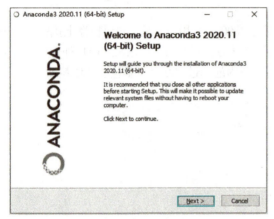

图 2-4 Anaconda 安装欢迎界面　　　　　图 2-5 Anaconda 最终用户许可协议

5）选择"Just Me（recommended）"单选按钮，表示能够执行 Anaconda 版本的用户只能是本人，这也是系统推荐的方式。类型选择之后，单击"Next"按钮，如图 2-6 所示。

6）单击"Browse"按钮后，选择安装 Anaconda 的文件路径。如果不自定义文件安装路径，系统将使用默认安装路径。路径设置完毕，单击"Next"按钮，如图 2-7 所示。

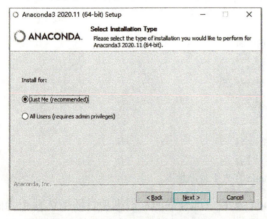
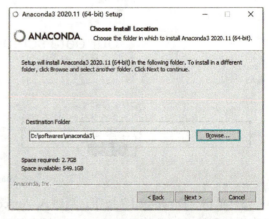

图 2-6 选择 Anaconda 安装类型　　　　　图 2-7 选择 Anaconda 安装路径

7)如果之前已经安装过其他版本的 Python,这里可以先不用勾选 "Add Anaconda3 to my PATH environment variable"复选框,可以在 Anaconda 安装完成之后手动完成环境变量的配置,并可以直接将原来安装 Python 的整个文件夹复制到 Anaconda 的 envs 目录下,实现由 Anaconda 进行统一管理。这里勾选 "Register Anaconda3 as my default Python 3.8"复选框,除非计划安装和运行多个版本的 Anaconda 或者多个版本的 Python,否则勾选此复选框,然后单击"Install"按钮,如图 2-8 所示。

8)开始安装 Anaconda,如图 2-9 所示。

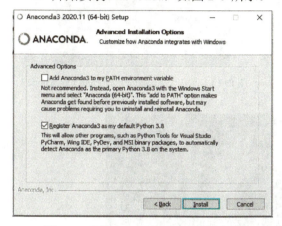 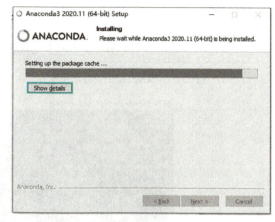

图 2-8 配置 Anaconda 安装环境变量　　　　　图 2-9 开始安装 Anaconda

9)安装部分完成,单击"Next"按钮,如图 2-10 所示。

10)此处是建议用户安装 PyCharm,并提供了下载链接:https://www.anaconda.com/pycharm。如果不安装,则单击"Next"按钮,如图 2-11 所示。

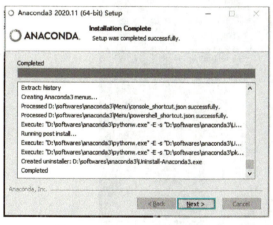 

图 2-10 Anaconda 安装部分完成　　　　　图 2-11 PyCharm 安装链接

11)Anaconda 安装结束,单击"Finish"按钮,如图 2-12 所示。

12)安装完成后,单击 Windows "开始"按钮,选择"Anaconda3(64-bit)",单击"Anaconda Prompt(anaconda3)"打开 Anaconda 命令行,如图 2-13 所示。

13)在 Anaconda Prompt(anaconda3)命令提示符中输入 conda list 就可以查询现在安装了哪些包,如图 2-14 所示。

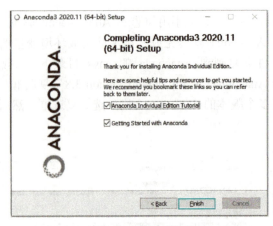

图 2-12　Anaconda 安装结束

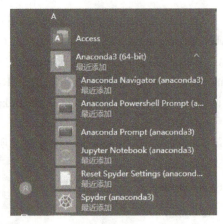

图 2-13　单击"Anaconda Prompt（anaconda3）"

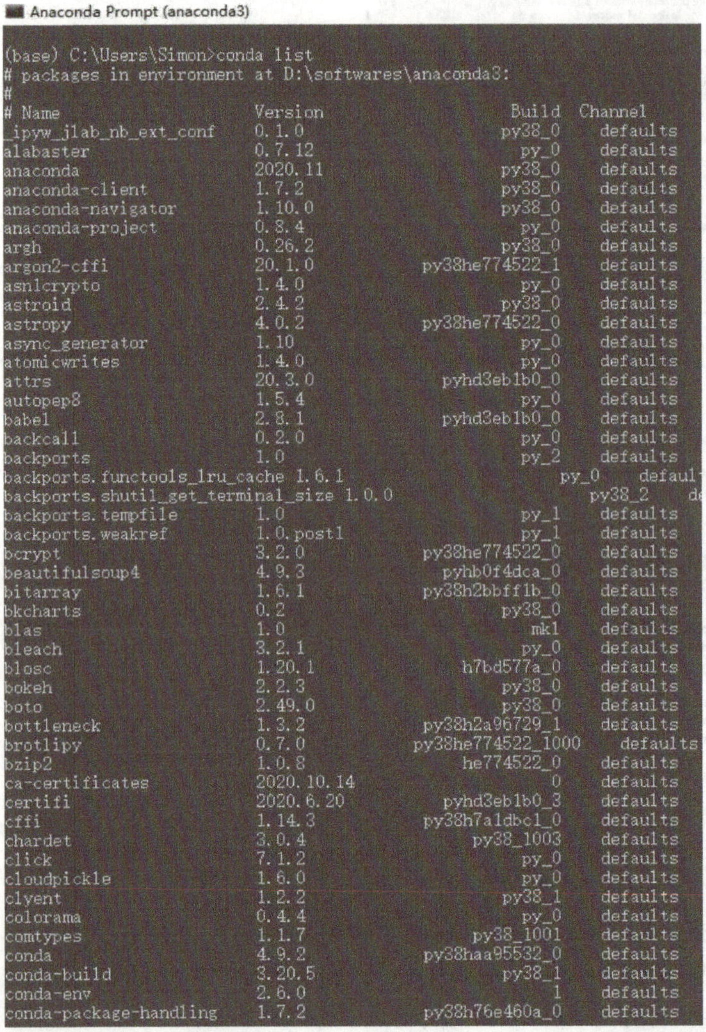

图 2-14　Anaconda 包查询

14）在 Anaconda Prompt（anaconda3）命令提示符中输入 conda info 可以查看 Anaconda 的基础环境，如图 2-15 所示。

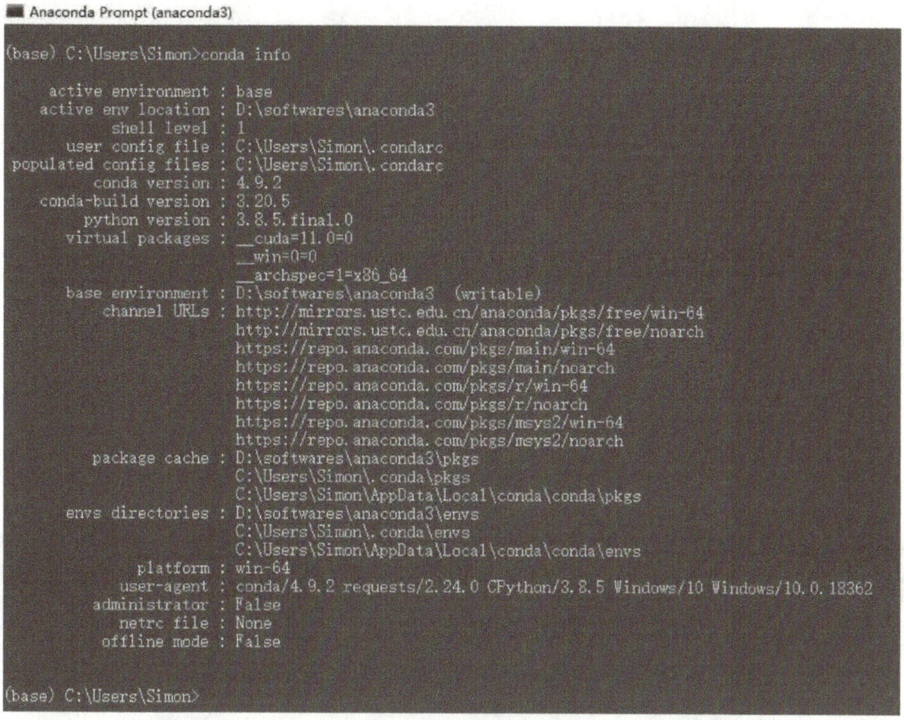

图 2-15 Anaconda 的基础环境

15）如果需要安装其他包，可以运行 conda install（包名）来进行安装。如果某个包版本不是最新的，运行 conda update（包名）就可以更新了。这里更新了 Pandas 包，如图 2-16 所示。

图 2-16 Anaconda 的 pandas 包更新

任务 2　运行 Jupyter Notebook

Jupyter Notebook 是通过使用浏览器网页形式编写和运行代码的工具，可以实现开发、文档编写、代码运行和结果展示。后面的数据分析案例将部分使用 Jupyter Notebook 实现。

1）Jupyter Notebook 包含在 Anaconda 包中。单击 Windows "开始" 按钮，选择 Anaconda 包中的 "Jupyter Notebook（anaconda3）" 即可，如图 2-17 所示。

2）单击 "Jupyter Notebook" 按钮之后就可以进入浏览器的操作界面了。这里显示的是当前 Jupyter Notebook 的文件夹和文件，如图 2-18 所示。

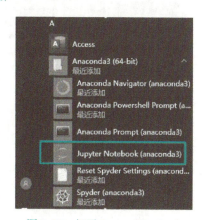

图 2-17　打开 Jupyter Notebook

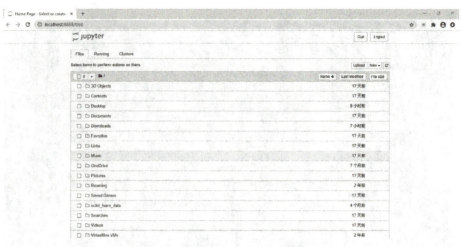

图 2-18　Jupyter Notebook 操作界面

3）单击"New"按钮，可以选择"Python 3"打开 Python 编辑器，选择"Text File"打开文本编辑器，选择"Folder"新建文件夹，选择"Terminal"打开 shell 终端。这里单击"Python 3"进入 Python 编辑器，如图 2-19 和图 2-20 所示。

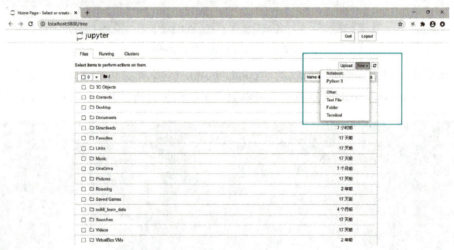

图 2-19　Jupyter Notebook 运行界面

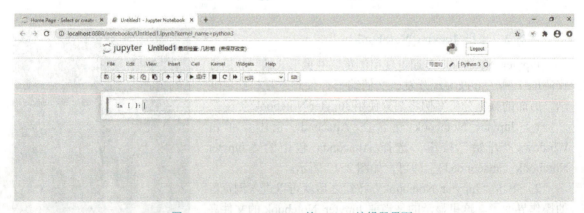

图 2-20　Jupyter Notebook 的 Python 编辑器界面

任务 3 PyCharm 的安装和使用

（1）PyCharm 简介

PyCharm 是用于 Python 语言开发的集成开发环境，具有跨平台性。PyCharm 具有十分强大的功能，能够帮助用户极大地提高开发效率，具体功能如下。

1）代码补全。

PyCharm 拥有代码补全功能，支持代码展开和折叠，能够非常高效地提示开发人员完成编码。

2）代码索引。

PyCharm 提供代码索引导航功能，能够协助开发人员快速、准确地找到文件、方法或者类之间声明和引用关系的位置。

3）代码高亮提示。

根据 Python 语言语法树规则，PyCharm 提供了语法错误高亮提示功能，能够及时检测项目代码并智能补全。

4）项目重构。

PyCharm 可以根据用户需求更改项目名称，获取方法和超类，导入域、变量和常量等。

5）支持主流的 Web 框架。

PyCharm 提供了能够编辑前端代码 HTML、CSS 和 JavaScript 的编辑器，通过 jinja2 实现与 Web 框架 Flask 和 Django 的数据交互，高效构建 Web 网站。

6）强大代码调试。

PyCharm 拥有非常优秀的代码调试器，通过设置断点、步进、分析与调优、多视图显示窗口和反馈调试信息等能够对项目代码进行开发调试和单元测试。

（2）PyCharm 安装

1）进入 PyCharm 官方网站https://www.jetbrains.com/pycharm/，如图 2-21 所示。

图 2-21 进入 PyCharm 官方网站

2）单击"Download"按钮之后，根据项目环境选择正确的操作系统环境。Professional 是专业版，Community 是社区版，推荐安装社区版，因为社区版是免费使用的。这里选择了基于 Windows 操作系统的 Community 免费社区版，如图 2-22 所示。

3）下载 PyCharm 安装包之后，双击该安装包进入安装欢迎界面，单击"Next"按钮，如图 2-23 所示。

4）单击"Browse"按钮选择 PyCharm 的安装路径。然后单击"Next"按钮，如图 2-24 所示。

图 2-22　选择 PyCharm 安装环境

图 2-23　PyCharm 安装欢迎界面

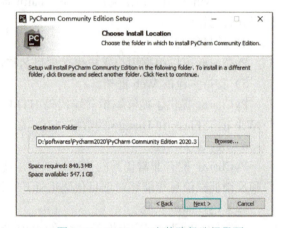

图 2-24　PyCharm 安装路径选择界面

5）勾选 64-bit launcher（64 位启动器）复选框，并创建.py 文件关联，在更新路径变量中勾选 64 位启动器目录到指定路径，在更新上下文菜单中勾选 Add "Open Folder as Project（添加文件夹作为项目）"。然后单击 "Next" 按钮，如图 2-25 所示。

6）将 "开始" 文件夹名自定义为 "JetBrains"，然后单击 "Install" 按钮，如图 2-26 所示。

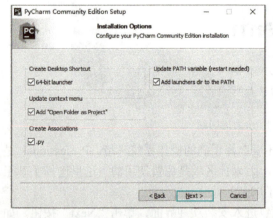

图 2-25　配置 PyCharm

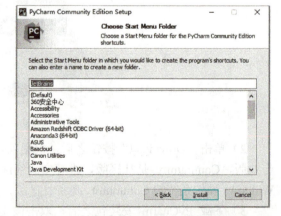

图 2-26　配置 PyCharm "开始" 菜单

7）安装完成后，单击 "Finish" 按钮。如果需要现在重新启动，则选中 "Reboot now" 单选按钮，如图 2-27 所示。

（3）PyCharm 运行

1）PyCharm 创建项目。首先，双击桌面的 PyCharm 图标打开 PyCharm 程序，选择 "File" 菜单→"New Project…" 选项，如图 2-28 所示。

2）在弹出窗口中的 "Location" 文本框中自定义项目名称为 test，并将该项目存放在指定位置 D:\pythonprojects\test。下面的 Python Interpreter: Python 3.8（base）中，可以选择新建一个虚拟环境或者使用一个之前已经存在的项目解释器。不同的项目可能会使用不同的依赖项，因此会使用

图 2-27　PyCharm 安装成功界面

不同的项目解释器。这里使用之前安装的 Anaconda 自带的解释器，开发人员可以根据实际需要确定项目解释器，如图 2-29 所示。

图 2-28　PyCharm 创建项目界面

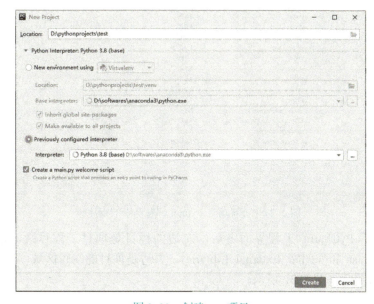

图 2-29　创建 test 项目

3）成功创建项目之后，建议首先查看一下本项目解释器所使用的各种包。这里可以首先选择"File"菜单→"Settings"选项，如图 2-30 所示。

图 2-30　进入 PyCharm 项目设置

4）进入"Settings"窗口之后，选择"Project test"→"Python Interpreter"菜单命令可以查看当前项目使用的 Python 解释器中所包含的程序包，如图 2-31 所示。

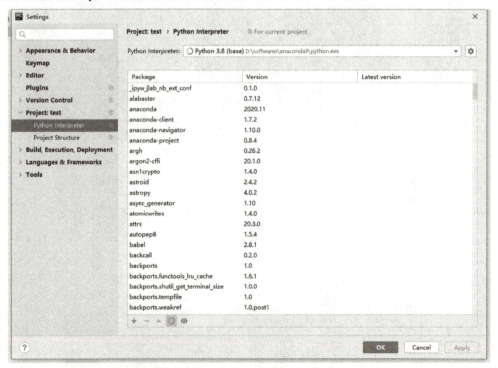

图 2-31　查看项目 Python 解释器中的程序包

5）在进入了 PyCharm 工程界面之后，左边的栏目是项目工程目录，包括项目根节点 D:\pythonprojects\test 和外部库 External Libraries。右边是项目的编辑区域，这里显示的是一个 main.py 的样例程序，如图 2-32 所示。

项目 2　安装 Python 数据分析工具

图 2-32　PyCharm 工程界面

6）将鼠标移到项目根节点，右击鼠标，选择"New"→"Python File"选项。这样就可以在 PyCharm 中创建一个基于 Python 的文件。在此，将该文件命名为 demo1，如图 2-33 所示。

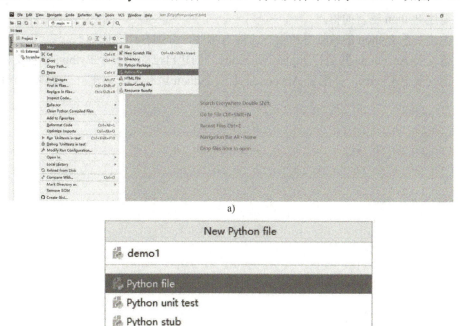

图 2-33　创建 Python 文件
a) 创建 Python 文件过程　b) 给 Python 文件命名

7）单击界面左下角的"Terminal"，可以在 PyCharm 中打开 CMD 命令行。输入"python"后就可以打开 CMD 中的 Python 解释器，如图 2-34 所示。

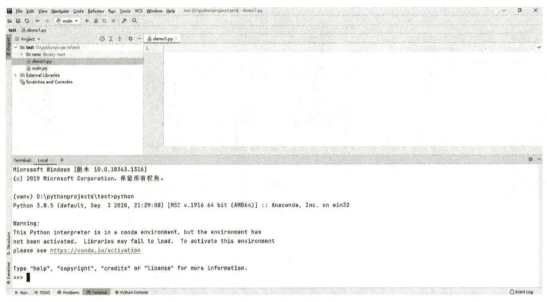

图 2-34　在 Terminal 中开启 Python 解释器

练习题

（1）简述 Python 语言的特点。

（2）简述基于 Python 的数据分析第三方库有哪些，分别有什么作用？

（3）请安装 Anaconda 并运行 Jupter Notebook，创建自己的第一个 Python 工程文档。

（4）请安装 PyCharm，并配置 Anaconda 中的 Python 解释器，创建自己的第一个 Python 工程文档。

项目 3　使用 NumPy 实现统计分析和处理

【项目分析】

本项目旨在理解和掌握 NumPy 数据分析工具的基本概念和基础语法。具体内容如下。
1）ndarry 对象的特点和内部结构以及使用实例。
2）NumPy 数据类型和应用 dtype 类实现数据类型的封装实例。
3）NumPy 数组和属性以及一维和多维数组及其属性的实例。
4）NumPy 切片和索引以及使用 slice 函数和数组下标索引的实例。
5）NumPy 常用函数在处理字符串、数学、算术、统计、排序和条件选择时的实例。
6）NumPy 数组维度操作函数的实例。
7）NumPy 创建数组函数的实例。
8）NumPy 常用 IO 函数的实例以及 NumPy 广播机制和实例。
9）NumPy 数据分析案例介绍。

【知识准备】

3.1　NumPy 的基本概念

3.1.1　NumPy 基础理论和引用方法

NumPy 支持创建 N 维数组和矩阵运算，并能够很好地支持 N 维数组的各种函数，极大地丰富和完善了 Python 的程序包。

学习并安装过 Python 的读者会发现 Python 官网发布的版本中是不包含 NumPy 的。因此，这里可以使用前面已经安装的 Anaconda。作为免费的 Python 发行版，Anaconda 不仅包含了常用的数据处理和分析等程序包，还能对这些程序包进行有效、统一的管理和部署，很好地免去了工作人员对纷繁复杂且相互依赖的 Python 程序包进行日常维护和管理的工作，从而让他们能够更好地专注于业务本身。这里通过引用 Anaconda 自带的 Python 解释器就可以导入 NumPy 包，如图 3-1 所示。

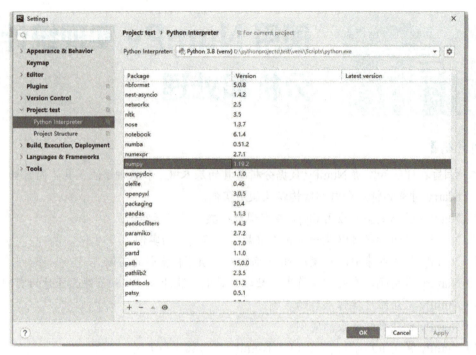

图 3-1　导入 NumPy 包

3.1.2　ndarry 对象

ndarray 对象是 NumPy 的重要组成元素。ndarray 对象能够创建 N 维数组对象，并通过其丰富的函数进行操作。ndarray 对象的特点如下。

1）ndarray 对象是以 0 为数组下标起点进行索引的多维数组。

2）ndarray 对象中存放的是同一类型数据。

3）ndarray 对象的每个元素拥有相同大小的内存空间。

4）ndarray 对象的内部组织结构包含 4 个部分：数据指针（Pointer）、描述数组的形状（Shape）元组、实现数组前后索引的步长（Stride）元组和描述元素的数据类型（Data-type）。

① 数据指针：用于指向数据所在内存空间的地址变量。

② 数组形状：用于描述数据维度信息的元组。

③ 数组索引步长：用于记录数组前后索引距离的元组。

④ 数据类型：用于表示数组内元素的数据属性。

ndarray 对象的内部结构如图 3-2 所示。

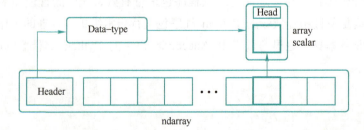

图 3-2　ndarray 对象的内部结构

【实例 3-1】 使用 NumPy 的 array 函数创建两个数组对象，即一个一维数组和一个多维数组。

```
#导入 NumPy 包，并命名为 np
import numpy as np
#使用 NumPy 的数组函数 array 创建一个拥有 3 个元素的整型数组
a = np.array([1,2,3])
print(a)
print('使用 NumPy 的数组函数 array 创建一个拥有 3 个元素的整型数组')
#使用 NumPy 的数组函数 array 创建多维数组
b = np.array([[1, 2], [3, 4]])
print (b)
print('使用 NumPy 的数组函数 array 创建多维数组')
```

输出如下：

[1 2 3]
使用 NumPy 的数组函数 array 创建一个拥有 3 个元素的整型数组
[[1 2]
 [3 4]]
使用 NumPy 的数组函数 array 创建多维数组

3.1.3 NumPy 数据类型

NumPy 拥有丰富的数据类型，应用 dtype 类实现数据类型的封装。NumPy 基本类型见表 3-1。

表 3-1　NumPy 基本类型

类型	描述	类型	描述
bool_	布尔型数据类型（True 或者 False）	uint8、uint16、uint32、uint64	无符号整数
int_	整数类型	int16、int32、int64	整数
intc	int32 或 int64	float16	半精度浮点数
intp	用于索引的整数类型	float32	单精度浮点数
int8	字节	float64	双精度浮点数
complex_	complex128 类型的另一种表示	float_	float64 类型的另一种表示
complex64	复数，表示双 32 位浮点数	complex128	复数，表示双 64 位浮点数

【实例 3-2】 使用 dtype 类创建数据类型，并赋予指定字段名。

```
import numpy as np
# 使用 dtype 类创建 int32 数据类型
a = np.dtype(np.int32)
# 使用 dtype 类创建 int8 数据类型，字段名为 age 的数据类型
b = np.dtype([('degree',np.int8)])
# 将数据类型应用于 ndarray 对象
c = np.array([(10,),(20,),(30,)], dtype = b)
print(a)
print('使用 dtype 类创建 int32 数据类型')
print(b)
print('使用 dtype 类创建 int8 数据类型，字段名为 degree 的数据类型')
```

```
print(c)
print('将数据类型应用于 ndarray 对象')
# 输出 ndarray 对象 c 中字段名为 degree 的值
print(c['degree'])
print('输出 ndarray 对象 c 中字段名为 degree 的值')
```

输出如下:

```
int32
使用 dtype 类创建 int32 数据类型
[('degree', 'i1')]
使用 dtype 类创建 int8 数据类型,字段名为 degree 的数据类型
[(10,) (20,) (30,)]
将数据类型应用于 ndarray 对象
[10 20 30]
输出 ndarray 对象 c 中字段名为 degree 的值
```

3.1.4 NumPy 数组属性

ndarray 拥有多个数组属性,其中主要的属性见表 3-2。

表 3-2 NumPy 数组属性

属性名称	描述	属性名称	描述
ndarray.ndim	表示数组的维度	ndarray.itemsize	数组中每个元素的字节大小
ndarray.shape	表示矩阵的行和列数	ndarray.flags	数组的内存信息
ndarray.size	数组元素的总个数	ndarray.real	数组元素的实部
ndarray.dtype	数组元素类型	ndarray.imag	数组元素的虚部

【实例 3-3】 创建一维数组 a 和多维数组 b,并输出 ndarray 的 ndim、shape、itemsize 和 flags 属性值。

```
import numpy as np
# 使用 arange 函数创建一个 0~39 的一维数组
a = np.arange(40)
# 输出 ndarray 数组 a 的 ndim 属性值,表示其维度数
print(a.ndim)
print('输出 ndarray 数组 a 的 ndim 属性值,表示其维度数')
# 使用 reshape 函数将一维的 ndarray 数组 a 重塑为一个三维数组 b
b = a.reshape(2,4,5)
# 输出 ndarray 数组 b 的 ndim 属性值,表示其维度数
print(b.ndim)
print('输出 ndarray 数组 b 的 ndim 属性值,表示其维度数')
# 输出 ndarray 数组 b 的 shape 属性值,表示其行列数
print (b.shape)
print('输出 ndarray 数组 b 的 shape 属性值,表示其行列数')
# 输出 ndarray 数组 b 的 itemsize 属性值,表示其内部每个元素字节大小
print (b.itemsize)
print('输出 ndarray 数组 b 的 itemsize 属性值,表示其内部每个元素字节大小')
# 输出 ndarray 数组 b 的 flags 属性值,表示其数组内存信息
print (b.flags)
print('输出 ndarray 数组 b 的 flags 属性值,表示其数组内存信息')
```

```
# 输出 ndarray 数组 b 的 size 属性值，表示其数组元素个数
print (b.size)
print('输出 ndarray 数组 b 的 size 属性值，表示其数组元素个数')
```

输出如下：

```
1
输出 ndarray 数组 a 的 ndim 属性值，表示其维度数
3
输出 ndarray 数组 b 的 ndim 属性值，表示其维度数
(2, 4, 5)
输出 ndarray 数组 b 的 shape 属性值，表示其行列数
4
输出 ndarray 数组 b 的 itemsize 属性值，表示其内部每个元素字节大小
  C_CONTIGUOUS : True
  F_CONTIGUOUS : False
  OWNDATA : False
  WRITEABLE : True
  ALIGNED : True
  WRITEBACKIFCOPY : False
  UPDATEIFCOPY : False
输出 ndarray 数组 b 的 flags 属性值，表示其数组内存信息
40
输出 ndarray 数组 b 的 size 属性值，表示其数组元素个数
```

3.1.5 NumPy 切片和索引

与 Python 的列表数组对象类似，ndarray 数组对象也是使用数组下标或者切片分割函数 slice()来索引和修改数据，并能够灵活地设置数组序列中的开始下标、结束下标和步长。

【实例 3-4】 使用 slice 函数对 ndarray 数组进行索引操作，并对 ndarray 数组使用切片操作。

```
import numpy as np
# 使用 arange 函数创建一个 0～9 的一维数组
a = np.arange(10)
# 使用 array 函数创建一个二维数组 b
c = np.array([[1,2,3,4],[4,5,6,7],[7,8,9,10]])
# 使用 slice 函数索引从第 3 个元素开始到第 8 个元素停止的元素，步长为 2
s = slice(2, 7, 2)
print(a[s])
print('使用 slice 函数索引从第 3 个元素开始到第 8 个元素停止的元素，步长为 2')
# 输出 ndarray 数组 a 中第 4 个元素的值
b = a[3]
print(b)
print('输出 ndarray 数组 a 中第 4 个元素的值')
# 输出 ndarray 数组 a 中从第 3 个元素开始到最后一个元素的值
print(a[2:])
print('输出 ndarray 数组 a 中从第 3 个元素开始到最后一个元素的值')
# 输出 ndarray 数组 a 中第 3 个元素到第 5 个元素的值
print(a[2:5])
print('输出 ndarray 数组 a 中第 3 个元素到第 5 个元素的值')
# 输出 ndarray 数组 a 中第 2 个元素到第 10 个元素,步长为 2 的值
print(a[1:10:2])
```

```
print('输出 ndarray 数组 a 中第 2 个元素到第 10 个元素,步长为 2 的值')
# 获取数组最后一个元素
print(a[-1])
print('获取数组最后一个元素')
# 获取数组第一个元素
print(a[-len(a)])
print('获取数组第一个元素')
# 获取二维数组中第 1 行 1、2 列的值
print(c[1,1:3])
print('获取二维数组中第 1 行 1、2 列的值')
# 获取二维数组中所有行列的值
print(c[:,:])
print('获取二维数组中所有行列的值')
# 获取二维数组中第 2 行到最后一行,倒数第 2 列到最后一列的值
print(c[1:-1,-2:-1])
print('获取二维数组中第 2 行到最后一行,倒数第 2 列到最后一列的值')
```

输出如下:

```
[2 4 6]
使用 slice 函数索引从第 3 个元素开始到第 8 个元素停止的元素,步长为 2
3
输出 ndarray 数组 a 中第 4 个元素的值
[2 3 4 5 6 7 8 9]
输出 ndarray 数组 a 中从第 3 个元素开始到最后一个元素的值
[2 3 4]
输出 ndarray 数组 a 中第 3 个元素到第 5 个元素的值
[1 3 5 7 9]
输出 ndarray 数组 a 中第 2 个元素到第 10 个元素,步长为 2 的值
9
获取数组最后一个元素
0
获取数组第一个元素
[5 6]
获取二维数组中第 1 行 1、2 列的值
[[ 1  2  3  4]
 [ 4  5  6  7]
 [ 7  8  9 10]]
获取二维数组中所有行列的值
[[6]]
获取二维数组中第 2 行到最后一行,倒数第 2 列到最后一列的值
```

3.2 NumPy 函数

3.2.1 NumPy 数学函数

NumPy 有着强大的数学函数集合,可以处理字符串、数学、算术、统计、排序和条件选择等数据,其中主要的数学函数见表 3-3。

表 3-3　NumPy 数学函数

函数名称	描　　述	函数名称	描　　述
add()	拼接数组元素	sort()	排序函数
splitlines()	按行进行分割	add()	加法函数
join()	指定分隔符来连接数组中的元素	subtract()	减法函数
strip()	删除字符串两端指定字符	multiply()	乘法函数
capitalize()	将字符串第一个字母转换为大写	divide()	除法函数
replace()	替换字符串中指定的内容	min()	求最小值
lower()	将字符串转为小写	max()	求最大值
split()	指定分隔符对字符串进行分割	sum()	求和
title()	每个单词的首字母转为大写	std()	求标准差
upper()	将字符串转为大写	var()	求数组方差
sin()	正弦函数	cumsum()	求数组元素累计和
cos()	余弦函数	cumprod()	求数组元素累计积
tan()	正切函数	repeat()	重复数组元素

【实例 3-5】　使用字符数组类 numpy.char 的 add()、multiply()、capitalize()、title()、lower()、upper()、split()、splitlines()、strip()、join()、replace()函数实现字符串的操作。

```
import numpy as np
print('字符串操作：')
# 使用 add 函数拼接两个字符串
print(np.char.add(['hello'], [' world']))
print('使用 add 函数拼接两个字符串')
# 使用 multiply 函数实现字符串多重拼接
print (np.char.multiply('hello ',3))
print('使用 multiply 函数实现字符串多重拼接')
# 使用 capitalize 函数将字符串首字母大写
print (np.char.capitalize('hello'))
print('使用 capitalize 函数将字符串首字母大写')
# 使用 title 函数将每一个单词首字母大写
print (np.char.title('i like china'))
print('使用 title 函数将每一个单词首字母大写')
# 使用 lower 函数将字符串全部小写
print (np.char.lower('HELLO'))
print('使用 lower 函数将字符串全部小写')
# 使用 upper 函数将字符串全部大写
print (np.char.upper('hello'))
print('使用 upper 函数将字符串全部大写')
# 使用 split 函数，设定分隔符为'.'，分割字符串
print (np.char.split ('www.hello.com', sep = '.'))
print('使用 split 函数，设定分隔符为'.'，分割字符串')
# 使用 splitlines 函数根据行分隔符'\n'对字符串按行进行分割
print (np.char.splitlines('i\nlike China'))
print('''使用 splitlines 函数根据行分隔符'\n'对字符串按行进行分割''')
# 使用 strip 函数除去字符串头尾的指定字符'a'
print (np.char.strip('ahello worlda','a'))
print('''使用 strip 函数除去字符串头尾的指定字符'a'''')
# 使用 join 函数将指定连接符':'插入字符串
```

```python
print (np.char.join(':','hello'))
print('使用join函数将指定连接符':'插入字符串''')
# 使用replace函数替换字符串中指定的内容
print (np.char.replace ('i like bireddata', 'red', 'g'))
print('使用replace函数替换字符串中指定的内容')
```

输出如下：
字符串操作：

```
['hello world']
```
使用add函数拼接两个字符串
```
hello hello hello
```
使用multiply函数实现字符串多重拼接
```
Hello
```
使用capitalize函数将字符串首字母大写
```
I Like China
```
使用title函数将每一个单词首字母大写
```
hello
```
使用lower函数将字符串全部小写
```
HELLO
```
使用upper函数将字符串全部大写
```
['www', 'hello', 'com']
```
使用split函数，设定分隔符为'.'，分割字符串
```
['i', 'like China']
```
使用splitlines函数根据行分隔符'\n'对字符串行进行分割
```
hello world
```
使用strip函数除去字符串头尾的指定字符'a'
```
h:e:l:l:o
```
使用join函数将指定连接符':'插入字符串
```
i like bigdata
```
使用replace函数替换字符串中指定的内容

【实例3-6】 使用sin()、cos()、tan()数学函数对数值进行基本操作。

```python
import numpy as np
a = np.array([0, 45, 90])
# 不同角度的正弦值
# 通过乘 pi/180 转化为弧度
print(np.sin(a * np.pi / 180))
print('正弦值')
# 数组中角度的余弦值
print(np.cos(a * np.pi / 180))
print('余弦值')
# 数组中角度的正切值
print(np.tan(a * np.pi / 180))
print('正切值')
```

输出如下：
```
[0.         0.70710678 1.        ]
```
正弦值
```
[1.00000000e+00 7.07106781e-01 6.12323400e-17]
```
余弦值
```
[0.00000000e+00 1.00000000e+00 1.63312394e+16]
```
正切值

【实例 3-7】 使用 add()、subtract()、multiply()、divide()算术函数对数值进行基本操作。

```
import numpy as np
b = np.array([200,400,800])
c = np.array([1,2,4])
d = np.array([[1,2,4],[8,10,20]])
# 两个数组相加
print (np.add(b,c))
print('b,c 两个数组相加')
print (np.add(b,d))
print('b,d 两个数组相加')
# 两个数组相减
print (np.subtract(b,c))
print('b,c 两个数组相减')
print (np.subtract(b,d))
print('b,d 两个数组相减')
# 两个数组相乘
print (np.multiply(b,c))
print('b,c 两个数组相乘')
print (np.multiply(b,d))
print('b,d 两个数组相乘')
# 两个数组相除
print (np.divide(b,c))
print('b,c 两个数组相除')
print (np.divide(b,d))
print('b,d 两个数组相除')
```

输出如下：

```
[201 402 804]
b,c 两个数组相加
[[201 402 804]
 [208 410 820]]
b,d 两个数组相加
[199 398 796]
b,c 两个数组相减
[[199 398 796]
 [192 390 780]]
b,d 两个数组相减
[ 200  800 3200]
b,c 两个数组相乘
[[  200   800  3200]
 [ 1600  4000 16000]]
b,d 两个数组相乘
[200. 200. 200.]
b,c 两个数组相除
[[200. 200. 200.]
 [ 25.  40.  40.]]
b,d 两个数组相除
```

【实例 3-8】 使用 max()、min()、std()、var()、cumsum()、cumprod()和 repeat()函数对数值进行基本操作。

```
# -*- coding: utf-8 -*-
import numpy as np
```

```
a = np.arange(1,10)
# 使用 max 函数求数组最大值
print(np.max(a))
print('使用 max 函数求数组最大值')
# 使用 min 函数求数组最小值
print(np.min(a))
print('使用 min 函数求数组最小值')
# 使用 std 函数求数组标准差
print(np.std(a))
print('使用 std 函数求数组标准差')
# 使用 var 函数求数组方差
print(np.var(a))
print('使用 var 函数求数组方差')
# 使用 cumsum 函数求数组元素累计和
print(np.cumsum(a))
print('使用 cumsum 函数求数组元素累计和')
# 使用 cumprod 函数求数组元素累计积
print(np.cumprod(a))
print('使用 cumprod 函数求数组元素累计积')
# 使用 repeat 函数重复数组元素两次
print(a.repeat(2))
print('使用 repeat 函数重复数组元素两次')
```

输出如下：

```
9
使用 max 函数求数组最大值
1
使用 min 函数求数组最小值
2.581988897471611
使用 std 函数求数组标准差
6.666666666666667
使用 var 函数求数组方差
[ 1  3  6 10 15 21 28 36 45]
使用 cumsum 函数求数组元素累计和
[     1      2      6     24    120    720   5040  40320 362880]
使用 cumprod 函数求数组元素累计积
[1 1 2 2 3 3 4 4 5 5 6 6 7 7 8 8 9 9]
使用 repeat 函数重复数组元素两次
```

3.2.2 NumPy 数组维度操作函数

当需要对数组进行进一步变化维度的操作时，NumPy 提供了强大的函数实现数组维度的操作，见表 3-4。

【实例 3-9】 使用 reshape()、ravel()、flatten()、hstack()、vstack()、concatenate()、hsplit()、vsplit() 和 split() 函数对数组维度进行基本操作。

```
# -*- coding: utf-8 -*-
import numpy as np
```

表 3-4 NumPy 数组维度操作函数

函数名称	描述
reshape()	重塑数组维度
ravel()	横向扁平多维数组为一维数组
flatten()	横向或纵向扁平多维数组为一维数组
hstack()	横向合并数组
vstack()	纵向合并数组
concatenate()	横向或纵向合并数组
hsplit()	横向分割数组
vsplit()	纵向分割数组
split()	横向或纵向分割数组

```python
# reshape()、ravel()、flatten()、hstack()、vstack()、concatenate()、
hsplit()、vsplit()和split()
# 创建一个二维数组 a
a = np.array([[0, 0, 0, 0],[1, 1, 1, 1],[2, 2, 2, 2],[3,3,3,3]])
# 创建一个二维数组 b
b = np.array([[1, 2, 3,4],[5, 6, 7,8],[9, 10, 11,12],[13, 14, 15,16]])
# 创建一个一维数组 c
c = np.array([1,2,3,4,5,6])

# 使用 reshape 将一维数组 c，修改为 2 行 3 列的二维数组
print(c.reshape(2,3))
print('使用 reshape 将一维数组 c，修改为 2 行 3 列的二维数组')

# 使用 ravel 将二维数组 a 横向扁平化为一维数组
print(a.ravel())
print('使用 ravel 将二维数组 a 横向扁平化为一维数组')

# 使用 flatten 将二维数组 a 纵向扁平化为一维数组
print(a.flatten('F'))
print('使用 flatten 将二维数组 a 纵向扁平化为一维数组')

# 使用 flatten 将二维数组 a 横向扁平化为一维数组
print(a.flatten())
print('使用 flatten 将二维数组 a 横向扁平化为一维数组')

# 使用 hstack 将二维数组 a 和 b 进行横向组合
print(np.hstack((a,b)))
print('使用 hstack 将二维数组 a 和 b 进行横向组合')

# 使用 vstack 将二维数组 a 和 b 进行纵向组合
print(np.vstack((a,b)))
print('使用 vstack 将二维数组 a 和 b 进行纵向组合')

# 使用 concatenate 将二维数组 a 和 b 进行横向组合
print(np.concatenate((a,b), axis=1))
print('使用 concatenate 将二维数组 a 和 b 进行横向组合')

# 使用 concatenate 将二维数组 a 和 b 进行纵向组合
print(np.concatenate((a,b), axis=0))
print('使用 concatenate 将二维数组 a 和 b 进行纵向组合')

# 使用 hsplit 将二维数组 a 横向分割成两部分
print(np.hsplit(a,2))
print('使用 hsplit 将二维数组 a 横向分割')

# 使用 vsplit 将二维数组 a 纵向分割成两部分
print(np.vsplit(a,2))
print('使用 vsplit 将二维数组 a 纵向分割')

# 使用 split 将二维数组 a 纵向分割成两部分
print(np.split(a,2,axis=0))
print('使用 split 将二维数组 a 纵向分割')

# 使用 split 将二维数组 a 横向分割成两部分
```

```
print(np.split(a,2,axis=1))
print('使用split将二维数组a横向分割')
```

输出如下：

```
[[1 2 3]
 [4 5 6]]
使用reshape将一维数组c，修改为2行3列的二维数组
[0 0 0 0 1 1 1 1 2 2 2 2 3 3 3 3]
使用ravel将二维数组a横向扁平化为一维数组
[0 1 2 3 0 1 2 3 0 1 2 3 0 1 2 3]
使用flatten将二维数组a纵向扁平化为一维数组
[0 0 0 0 1 1 1 1 2 2 2 2 3 3 3 3]
使用flatten将二维数组a横向扁平化为一维数组
[[ 0  0  0  0  1  2  3  4]
 [ 1  1  1  1  5  6  7  8]
 [ 2  2  2  2  9 10 11 12]
 [ 3  3  3  3 13 14 15 16]]
使用hstack将二维数组a和b进行横向组合
[[ 0  0  0  0]
 [ 1  1  1  1]
 [ 2  2  2  2]
 [ 3  3  3  3]
 [ 1  2  3  4]
 [ 5  6  7  8]
 [ 9 10 11 12]
 [13 14 15 16]]
使用vstack将二维数组a和b进行纵向组合
[[ 0  0  0  0  1  2  3  4]
 [ 1  1  1  1  5  6  7  8]
 [ 2  2  2  2  9 10 11 12]
 [ 3  3  3  3 13 14 15 16]]
使用concatenate将二维数组a和b进行横向组合
[[ 0  0  0  0]
 [ 1  1  1  1]
 [ 2  2  2  2]
 [ 3  3  3  3]
 [ 1  2  3  4]
 [ 5  6  7  8]
 [ 9 10 11 12]
 [13 14 15 16]]
使用concatenate将二维数组a和b进行纵向组合
[array([[0, 0],
       [1, 1],
       [2, 2],
       [3, 3]]), array([[0, 0],
       [1, 1],
       [2, 2],
       [3, 3]])]
使用hsplit将二维数组a横向分割
[array([[0, 0, 0, 0],
       [1, 1, 1, 1]]), array([[2, 2, 2, 2],
       [3, 3, 3, 3]])]
```

使用 vsplit 将二维数组 a 纵向分割
[array([[0, 0, 0, 0],
       [1, 1, 1, 1]]), array([[2, 2, 2, 2],
       [3, 3, 3, 3]])]
使用 split 将二维数组 a 纵向分割
[array([[0, 0],
       [1, 1],
       [2, 2],
       [3, 3]]), array([[0, 0],
       [1, 1],
       [2, 2],
       [3, 3]])]
使用 split 将二维数组 a 横向分割

3.2.3 NumPy 创建数组函数

NumPy 提供了丰富的内置函数，用于生成含有不同元素的数组对象，见表 3-5。

表 3-5 NumPy 创建数组函数

函数名称	描述
arrange()	生成指定范围的序列数组
zeros()	生成指定维度的数组，数组元素都为 0
ones()	生成指定维度的数组，数组元素都为 1
eye()	生成指定维度的数组，数组对角线元素都为 1，其余为 0
diag()	生成指定维度的数组，数组对角线元素都为指定元素值，其余为 0
linespace()	生成指定范围的等差数组
logspace()	生成指定范围的等比数组
random.random()	生成指定个数的随机数组
random.rand()	生成指定范围的随机数组
random.randn()	生成指定范围的正态分布的随机数组

【实例 3-10】 使用 NumPy 内置函数创建自定义数组。

```
# -*- coding: utf-8 -*-
import numpy as np

# 使用 arange 函数创建一个取值范围为 0～6 的一维数组,步长为 2
a = np.arange(0, 7, 2)
print(a)
print('# 使用 arange 函数创建一个取值范围为 0～6 的一维数组,步长为 2')

# 使用 zeros 函数创建一个 3 行 4 列的数组
print(np.zeros((3,4)))
print('使用 zeros 函数创建一个 3 行 4 列的数组')

# 使用 ones 函数创建一个 3 行 4 列的数组
print(np.ones((3,4)))
print('使用 ones 函数创建一个 3 行 4 列的数组')
```

```python
# 使用eye函数创建一个4行4列的数组，数组对角线元素为1
print(np.eye(4))
print('使用eye函数创建一个4行4列的数组，数组对角线元素为1')

# 使用diag函数创建一个4行4列的数组，数组对角线元素为a的元素值
print(np.diag(a))
print('使用diag函数创建一个4行4列的数组，数组对角线元素为a的元素值')

# 使用linespace函数创建取值范围0～10的5个等差数据
print(np.linspace(0,10,5))
print('使用linespace函数创建取值范围0～10之间的5个等差数据')

# 使用logspace函数创建取值范围0～10之间的5个等比数据
print(np.logspace(0,10,5))
print('使用logspace函数创建取值范围0～10之间的5个等比数据')

# 使用random.random函数生成10个一维随机数组
print(np.random.random(10))
print('使用random.random函数生成10个一维随机数组')

# 使用random.rand函数生成2行3列均匀分布的随机数组
print(np.random.rand(2,3))
print('使用random.rand函数生成2行3列均匀分布的随机数组')

# 使用random.randn函数生成2行3列正态分布的随机数组
print(np.random.randn(2,3))
print('使用random.randn函数生成2行3列正态分布的随机数组')
```

输出如下：

```
[0 2 4 6]
# 使用arange函数创建一个取值范围为0～6的一维数组，步长为2
[[0. 0. 0. 0.]
 [0. 0. 0. 0.]
 [0. 0. 0. 0.]]
使用zeros函数创建一个3行4列的数组
[[1. 1. 1. 1.]
 [1. 1. 1. 1.]
 [1. 1. 1. 1.]]
使用ones函数创建一个3行4列的数组
[[1. 0. 0. 0.]
 [0. 1. 0. 0.]
 [0. 0. 1. 0.]
 [0. 0. 0. 1.]]
使用eye函数创建一个4行4列的数组，数组对角线元素为1
[[0 0 0 0]
 [0 2 0 0]
 [0 0 4 0]
 [0 0 0 6]]
使用diag函数创建一个4行4列的数组，数组对角线元素为a的元素值
[ 0.  2.5  5.   7.5 10. ]
使用linespace函数创建取值范围0～10之间的5个等差数据
[1.00000000e+00 3.16227766e+02 1.00000000e+05 3.16227766e+07
```

```
1.00000000e+10]
```
使用 logspace 函数创建取值范围 0～10 之间的 5 个等比数据
```
[0.77896157 0.64005038 0.373311   0.8995392  0.8841037  0.39674863
 0.3728455  0.91932241 0.70121061 0.70761702]
```
使用 random.random 函数生成 10 个一维随机数组
```
[[0.98134146 0.60876466 0.24958598]
 [0.62934249 0.47094212 0.75246914]]
```
使用 random.rand 函数生成 2 行 3 列均匀分布的随机数组
```
[[-0.71390078 -0.27039802 -1.10526544]
 [-0.16046227  0.24969558 -0.37556257]]
```
使用 random.randn 函数生成 2 行 3 列正态分布的随机数组

3.2.4 NumPy 常用 IO 函数

NumPy 有多种格式文件读写的函数，可以保存和提取多种格式的文件数据，包括文本文件、二进制文件和 CSV 文件，见表 3-6。

表 3-6 NumPy 常用 IO 函数

函 数 名 称	描 述
tofile()	将数组中的数据以二进制格式写进文件
fromfile()	读回数据时需要用户指定元素类型，并对数组的形状进行适当的修改
save()	保存一个数组到一个文件中
savez()	保存多个数组到一个文件中
load()	读取文件
savetxt()	保存数据为文本格式，包括 CSV 格式的文件
loadtxt()	读取文本格式数据，包括 CSV 格式的文件

【实例 3-11】 使用 tofile()和 fromfile()保存和读取文件。

```
# -*- coding: utf-8 -*-
import numpy as np
# 使用 arange 和 reshape 函数创建一个 5 行 4 列的二维数组 a
a = np.arange(0, 20).reshape(5, 4)
# 使用 tofile 函数保存 a 为 a.bin
a.tofile('a.bin')
# 输出 a 的类型
print(a.dtype)
print('输出 a 的类型')
# 使用 fromfile 函数按照 int32 类型读入数据
b = np.fromfile('a.bin', dtype=np.int32)
# b 的数值是一维的
print(b)
print('b 的数值是一维的')
# 将一维数组 b 修改为 5 行 4 列
b.shape = 5, 4
print(b)
print('将一维数组 b 修改为 5 行 4 列')
```

输出如下：

iint32
输出 a 的类型
[0 1 2 3 4 5 6 7 8 9 10 11 12 13 14 15 16 17 18 19]
b 的数值是一维的
[[0 1 2 3]
 [4 5 6 7]
 [8 9 10 11]
 [12 13 14 15]
 [16 17 18 19]]
将一维数组 b 修改为 5 行 4 列

【实例 3-12】 使用 save()、savez()和 load()保存和读取文件。

```
# -*- coding: utf-8 -*-
import numpy as np
# 使用 arange 和 reshape 函数创建一个 5 行 4 列的数组 a
a = np.arange(0, 20).reshape(5, 4)
# 使用 save 函数将 a 保存为 a.npy。
np.save('a.npy', a)
# 使用 load 函数读取 a.npy
c = np.load('a.npy')
print(c)
print('使用 load 函数读取 a.npy')
# 使用 array 创建 2 行 3 列的二维数组 b
b = np.array([[3, 4, 5], [6, 7, 8]])
# 使用 arange 函数创建取值范围为 0~1, 步长为 0.1 的一维数组 c
c = np.arange(0, 1.0, 0.1)
# 使用 sin 函数返回一维数组 d
d = np.sin(c)
# 使用 savez 函数将数组 b、c 和 d 共同保存到文件 bcd.npz 中。
np.savez('bcd.npz', b, c, sin_array=d)
# 使用 load 函数读取 bcd.npz
r = np.load('bcd.npz')
# 数组 b
print(r['arr_0'])
print('数组 b')
# 数组 c
print(r['arr_1'])
print('数组 c')
# 数组 d
print(r['sin_array'])
print('数组 d')
```

输出如下:

[[0 1 2 3]
 [4 5 6 7]
 [8 9 10 11]
 [12 13 14 15]
 [16 17 18 19]]
使用 load 函数读取 a.npy
[[3 4 5]
 [6 7 8]]
数组 b

[0. 0.1 0.2 0.3 0.4 0.5 0.6 0.7 0.8 0.9]
数组 c
[0.   0.09983342 0.19866933 0.29552021 0.38941834 0.47942554
 0.56464247 0.64421769 0.71735609 0.78332691]
数组 d

【实例 3-13】 使用 savetxt() 和 loadtxt() 保存和读取文件。

```
# -*- coding: utf-8 -*-
import numpy as np
# 使用 arange 函数创建一个 4 行 4 列的二维数组 a
a = np.arange(0, 8, 0.5).reshape(4, 4)
# 使用 savetxt 函数保存二维数组 a 为文件 a.txt
np.savetxt('a.txt', a)
# 使用 loadtxt 函数读取文件 a.txt
b = np.loadtxt('a.txt')
print(b)
print('使用 loadtxt 函数读取文件 a.txt')

# 使用 savetxt 函数保存二维数组 a 的元素为整数，以逗号分割
np.savetxt('a.txt', a, fmt='%d', delimiter=',')
# 使用 loadtxt 函数加载文件 a.txt，并以逗号分割
c = np.loadtxt('a.txt', delimiter=',')
print(c)
print('使用 loadtxt 函数加载文件 a.txt，并以逗号分割')
# 使用 loadtxt 函数读取 test.csv 格式数据
tmp = np.loadtxt('test.csv', dtype=np.str, delimiter=',', encoding= 'utf-8')
print(tmp)
print('使用 loadtxt 函数读取 test.csv 格式数据')
# 读取数据，并转换数据为 float 型
data = tmp[1:, 1:].astype(np.float)
print(data)
print('读取数据，并转换数据为 float 型')
```

输出如下：
[[0. 0.5 1. 1.5]
 [2. 2.5 3. 3.5]
 [4. 4.5 5. 5.5]
 [6. 6.5 7. 7.5]]
使用 loadtxt 函数读取文件 a.txt
[[0. 0. 1. 1.]
 [2. 2. 3. 3.]
 [4. 4. 5. 5.]
 [6. 6. 7. 7.]]
使用 loadtxt 函数加载文件 a.txt，并以逗号分割
[['姓名' '年龄' '体重' '身高']
 ['Tom' '35' '85' '185']
 ['Simon' '65' '70' '170']
 ['Joe' '45' ' 60' '167']]
使用 loadtxt 函数读取 test.csv 格式数据
[[35. 85. 185.]
 [65. 70. 170.]
 [45. 60. 167.]]

读取数据，并转换数据为 float 型

3.2.5 NumPy 广播

广播（Broadcast）是 NumPy 对不同维度的数组进行数值计算的方式，对数组的算术运算通常在对应的元素上进行。广播的核心规则是：如果两个数组的拓展虚拟维度（纵轴或横轴）的轴长度相符，则认为它们是广播兼容的。广播会在虚拟的维度上进行计算，如图 3-3 和图 3-4 所示。

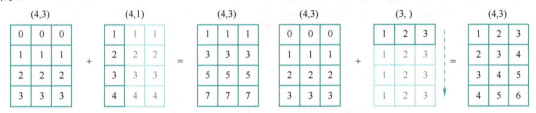

图 3-3　广播的规则 1　　　　　　　　图 3-4　广播的规则 2

【实例 3-14】使用 NumPy 的广播机制实现数组间的算术运算。

```
# -*- coding: utf-8 -*-
import numpy as np
# 创建 4 行 3 列的数组 a
a = np.array([[0, 0, 0],[1, 1, 1],[2, 2, 2], [3, 3, 3]])
# 创建 4 行 1 列的数组 b
b = np.array([[1],[2],[3],[4]])
# 创建 1 行 3 列的数组 c
c = np.array([1,2,3])
# 将 a 和 b 相加
ab1 = a + b
print(ab1)
print('将 a 和 b 相加')
# 将 a 和 b 相减
ab2 = a - b
print(ab2)
print('将 a 和 b 相减')
# 将 a 和 b 相乘
ab3 = a * b
print(ab3)
print('将 a 和 b 相乘')
# 将 a 和 b 相除
ab4 = a / b
print(ab4)
print('将 a 和 b 相除')
# 将 a 和 c 相加
ac1 = a + c
print(ac1)
print('将 a 和 c 相加')
# 将 a 和 c 相减
ac2 = a + c
print(ac2)
print('将 a 和 c 相减')
```

```
# 将 a 和 c 相乘
ac3 = a + c
print(ac3)
print('将 a 和 c 相乘')
# 将 a 和 c 相除
ac4 = a + c
print(ac4)
print('将 a 和 c 相除')
```

输出如下：
```
[[1 1 1]
 [3 3 3]
 [5 5 5]
 [7 7 7]]
将 a 和 b 相加
[[-1 -1 -1]
 [-1 -1 -1]
 [-1 -1 -1]
 [-1 -1 -1]]
将 a 和 b 相减
[[ 0  0  0]
 [ 2  2  2]
 [ 6  6  6]
 [12 12 12]]
将 a 和 b 相乘
[[0.         0.         0.        ]
 [0.5        0.5        0.5       ]
 [0.66666667 0.66666667 0.66666667]
 [0.75       0.75       0.75      ]]
将 a 和 b 相除
[[1 2 3]
 [2 3 4]
 [3 4 5]
 [4 5 6]]
将 a 和 c 相加
[[1 2 3]
 [2 3 4]
 [3 4 5]
 [4 5 6]]
将 a 和 c 相减
[[1 2 3]
 [2 3 4]
 [3 4 5]
 [4 5 6]]
将 a 和 c 相乘
[[1 2 3]
 [2 3 4]
 [3 4 5]
 [4 5 6]]
将 a 和 c 相除
```

【任务实施】

任务1 使用 NumPy 实现股票数据分析

利用数学和统计分析函数完成统计分析应用，包括计算平均价和加权平均价，最高价和最低价，最高差价和最低差价，开盘价的中位数和方差，收益率、年波动率和月波动率，股票数据如图3-5所示。

股票名称	日期	开盘价	最高价	最低价	收盘价	交易额
股票A	28-01-2021	244.17	244.4	233.53	236.1	17144800
股票A	31-01-2021	235.8	240.04	234.3	239.32	17443020
股票A	01-02-2021	241.3	245.65	240.98	245.03	12238002
股票A	02-02-2021	244.45	245.25	243.55	244.32	7249601
股票A	03-02-2021	243.8	244.24	238.55	243.44	12263120
股票A	04-02-2021	243.61	246.7	243.51	246.5	17424760
股票A	07-02-2021	247.89	253.25	247.64	251.88	11986300
股票A	08-02-2021	253.68	255.52	252.15	255.2	12029500
股票A	09-02-2021	255.19	259	254.87	258.16	17531590

图3-5 股票数据

1) 导入 NumPy 库，并命名为 np。

```
# -*- coding: utf-8 -*-
import numpy as np
```

2) dict()函数用于创建一个字典数据:dic_data1。

```
dic_data1 = dict(
    # fname 表示读取的文件
    fname = "股票分析数据.csv",
    # delimiter 表示数据分隔符
    delimiter = ',',
    # usecols 表示读取的列坐标
    usecols = (2,3,4,5,6),
    # skiprows 跳过前 x 行, 1 表示跳过第一行表头
    skiprows = 1,
    # encoding = 'utf-8'表示使用指定的字符集打开文件
    encoding = 'utf-8',
    # unpack 表示如果为 True,读入属性将分别写入不同数组变量, False 读入数据只写入一个数组变量，默认为 False
    unpack = True
)
```

3) 读取指定字段数据，并输出收盘价和交易额，如图3-6所示。

```
# 使用 np.loadtxt 方法读取字典 dic_data1,并将指定列的数据分别赋值给 col2、col3、col4、col5、col6
col2,col3,col4,col5,col6 = np.loadtxt(**dic_data1)
# 输出收盘价和交易额
print("收盘价=",col5)
print("交易额=",col6)
```

```
收盘价= [236.1  239.32 245.03 244.32 243.44 246.5  251.88 255.2  258.16]
交易额= [17144800. 17443020. 12238002.  7249601. 12263120. 17424760. 11986300.
 12029500. 17531590.]
```

图3-6 收盘价和交易额

4）使用 np.average 方法计算平均价和加权平均价，如图 3-7 所示。

```
col5_average = np.average(col5)
col6_wave = np.average(a,weights=col6)
print("平均价=",col5_average)
print("加权平均价=",col6_wave)
```

5）使用 np.max 和 np.min 方法计算股票最高价和最低价，如图 3-8 所示。

```
col3_maxPrice = np.max(col3)
col4_minPrice = np.min(col4)
print("最高价=",col3_maxPrice)
print("最低价=",col4_minPrice)
```

平均价= 246.6611111111111
加权平均价= 246.48958859384808

最高价= 259.0
最低价= 233.53

图 3-7　平均价和加权平均价　　　　　图 3-8　股票最高价和最低价

6）使用 np.ptp 方法计算股票最高差价和最低差价，如图 3-9 所示。

```
col3_highvalue = np.ptp(col3)
col4_lowvalue = np.ptp(col4)
print("最高差价=",col3_highvalue)
print("最低差价=",col4_lowvalue)
```

7）计算开盘价中位数和方差，如图 3-10 所示。

```
# 使用 np.median 方法计算开盘价的中位数
col2_midPrice = np.median(col2)
print("开盘价的中位数=",col2_midPrice)
# 使用 np.var 计算开盘价的方差
col2_var = np.var(col2)
print("开盘价的方差=",col2_var)
```

最高差价= 18.960000000000008
最低差价= 21.340000000000003

开盘价的中位数= 244.17
开盘价的方差= 31.95191111111107

图 3-9　股票最高差价和最低差价　　　　图 3-10　开盘价中位数和方差

8）计算收益率、年波动率和月波动率，如图 3-11 所示。

年波动率= 13.205983252844494
月波动率= 2.881781803927573

图 3-11　年波动率和月波动率

```
# 使用 np.diff 方法计算股票收益率
col5_logearning = np.diff(np.log(col5))
# 使用 std()函数计算方差，mean()函数计算均值，sqrt()函数计算平方根
col5_year_volatility = col5_logearning.std()/col5_logearning.mean()*np.sqrt(252)
col5_month_volatility = col5_logearning.std()/col5_logearning.mean()*np.sqrt(12)
# 输出年波动率和月波动率
print("年波动率=",col5_year_volatility)
print("月波动率=",col5_month_volatility)
```

任务 2　使用 NumPy 实现豆瓣电影数据分析

利用数学和统计分析函数完成统计分析应用，包括最高投票数量、最低投票数量、最长电影时长、最短电影时长、最高评分、最低评分及其对应的电影名称，电影数据（部分）如

图 3-12 所示。

A	B	C	D	E	F	G	H	I
	名字	投票人数	类型	产地	上映时间	时长	年代	评分
0	肖申克的救赎	692795	剧情/犯罪	美国	1994/9/10 0:00	142	1994	9.6
1	控方证人	42995	剧情/悬疑	美国	1957/12/17 0:00	116	1957	9.5
2	美丽人生	327855	剧情/喜剧	意大利	1997/12/20 0:00	116	1997	9.5
3	阿甘正传	580897	剧情/爱情	美国	1994/6/23 0:00	142	1994	9.4
4	霸王别姬	478523	剧情/爱情	中国	1993/1/1 0:00	171	1993	9.4
5	泰坦尼克号	157074	剧情/爱情	美国	2012/4/10 0:00	194	2012	9.4
6	辛德勒的名单	306904	剧情/历史	美国	1993/11/30 0:00	195	1993	9.4
7	新世纪福音战士剧场版：Air/	24355	剧情/动作	日本	1997/7/19 0:00	87	1997	9.4
8	银魂完结篇：直到永远的万事	21513	剧情/动画	日本	2013/7/6 0:00	110	2013	9.4
9	这个杀手不太冷	662552	剧情/动作	法国	1994/9/14 0:00	133	1994	9.4
10	灿烂人生	16807	剧情/爱情	意大利	2003/6/22 0:00	366	2003	9.3
11	疯狂动物城	284652	喜剧/动作	美国	2016/3/4 0:00	109	2016	9.3
12	福音战士新剧场版：破 ヱヴ	32524	剧情/动作	日本	2009/6/27 0:00	108	2009	9.3
13	海豚湾	159302	纪录片	美国	2009/7/31 0:00	92	2009	9.3
14	回忆积木小屋 つみきのい	93384	剧情/动画	日本	2008/6/10 0:00	12	2008	9.3
15	机器人总动员	421734	喜剧/爱情	美国	2008/6/27 0:00	98	2008	9.3
16	十二怒汉	134949	剧情	美国	1957/4/1 0:00	96	1957	9.3
17	旅行到宇宙边缘	10044	纪录片	美国	2008/11/7 0:00	90	2008	9.3
18	父与女	53358	剧情/动画	英国	2001/5/27 0:00	60	2001	9.2
19	暴力云与送子鹳	75567	喜剧/动画	美国	2009/5/29 0:00	6	2009	9.2
20	城市之光	31105	剧情/喜剧	美国	1931/1/30 0:00	87	1931	9.2
21	大闹天宫	74881	动画/奇幻	中国	1905/5/14 0:00	114	1961	9.2
22	盗梦空间	642134	剧情/动作	美国	2010/9/1 0:00	148	2010	9.2
23	放牛班的春天	370585	剧情/音乐	法国	2004/10/16 0:00	97	2004	9.2
24	海上钢琴师	501153	剧情/音乐	意大利	1998/10/28 0:00	165	1998	9.2
25	家园	32717	纪录片	法国	2009/6/5 0:00	118	2009	9.2
26	教父	280871	剧情/犯罪	美国	1972/3/15 0:00	175	1972	9.2
27	乱世佳人	226131	剧情/爱情	美国	1939/12/15 0:00	238	1939	9.2
28	千与千寻 千と千尋の神隠	525505	剧情/动画	日本	2001/7/20 0:00	125	2001	9.2
29	穹顶之下	51113	纪录片	中国	2015/2/28 0:00	104	2015	9.2
30	生活多美好	24495	剧情/爱情	美国	1946/12/20 0:00	130	1946	9.2
31	唐顿庄园：2011圣	29611	剧情	英国	2011/12/25 0:00	93	2011	9.2
32	(无)	331	剧情/喜剧	美国	2011/6/18 0:00	75	2011	7

图 3-12　电影数据（部分）

1）导入 NumPy 库，并命名为 np。

```
# -*- coding: utf-8 -*-
import numpy as np
```

2）使用 dict()函数创建一个字典数据:dic_data1。

```
dic_data1 = dict(
    # fname 表示读取的文件
    fname = "电影数据.csv",
    # delimiter 表示数据分隔符
    delimiter = ',',
    # usecols 表示读取的列坐标
    usecols = (2,6,8),
    # skiprows 跳过前 x 行，1 表示跳过第一行表头
    skiprows = 1,
    # encoding = 'utf-8'表示使用指定的字符集打开文件
    encoding = 'utf-8',
    # unpack 表示如果为 True,读入属性将分别写入不同数组变量，False 读入数据只写入一个
数组变量，默认 False
    unpack = True
    )
```

3）使用 np.loadtxt 方法读取字典 dic_data1，并将指定列的数据统一赋值给 a，组成一个新的 ndarray 数组，如图 3-13～图 3-15 所示。

```
a = np.loadtxt(**dic_data1)
# 通过 a 的数组下标分别获取指定的字段数据：投票数量、电影时长、电影评分
print("投票数量=",a[0])
```

```
print("电影时长=",a[1])
print("电影评分=",a[2])
```

投票数量= [6.92795e+05 4.29950e+04 3.27855e+05 5.80897e+05 4.78523e+05 1.57074e+05
 3.06904e+05 2.43550e+04 2.15130e+04 6.62552e+05 1.68070e+05 2.84652e+05
 3.25240e+04 1.59302e+05 9.33840e+04 4.21734e+05 1.34949e+05 1.00440e+04
 5.33580e+04 7.55670e+04 3.11050e+04 7.48810e+04 6.42134e+05 3.70585e+05
 5.01153e+05 3.27170e+04 2.80871e+05 2.26131e+05 5.25505e+05 5.11130e+04
 2.44950e+04 2.96110e+04 4.31000e+02 1.05300e+05 1.70590e+04 1.74930e+04
 3.52613e+05 1.32840e+04 1.06780e+04 1.37570e+04 1.09230e+05 5.40000e+01
 9.70000e+01 1.25840e+04 1.08230e+04 1.07810e+04 3.96000e+02 1.85620e+04
 3.80430e+04 1.09000e+02 1.34273e+05 2.93040e+05 1.24000e+02 3.66223e+05
 1.43000e+02 1.88550e+04 5.40000e+04 2.38600e+03 1.85000e+02 8.80000e+01
 8.10000e+01 8.00000e+01 2.82500e+03 6.92300e+03 3.51500e+04 2.87260e+04
 4.56820e+04 1.80738e+05 2.38320e+04 1.34400e+03 2.11100e+03 3.69900e+03
 5.42000e+03 3.55400e+03 4.16670e+03 3.78310e+04 4.21400e+03 3.90700e+03
 3.53400e+03 3.23800e+03 3.98100e+03 1.51333e+05 3.63100e+03 6.23600e+03
 4.09200e+03 3.77200e+03 4.60200e+03 3.44318e+05 3.96600e+03 1.41130e+05
 6.49160e+04 1.26000e+02 5.95000e+02 5.88450e+04 7.80000e+01 7.53000e+02
 4.57070e+04 1.76426e+05 1.66196e+05 5.49808e+05 1.46000e+02 1.86000e+02
 1.14819e+05 9.24000e+02 9.51400e+03 6.01000e+02 1.77000e+03 1.53100e+03
 1.49000e+02 5.35491e+05 1.80760e+04 2.59119e+05 1.88690e+04 9.83400e+03
 3.03460e+04 9.80700e+03 1.07031e+05 9.79700e+03 9.71100e+03 9.58300e+03]

图 3-13 投票数量

电影时长= [142. 116. 116. 142. 171. 194. 195. 87. 110. 133. 366. 109. 108. 92.
  12. 98. 96. 90. 60.   6. 87. 114. 148. 97. 165. 118. 175. 238.
 125. 104. 130. 93. 75. 14. 101. 96. 93. 48. 118. 132. 109. 49.
  89. 113. 93. 19. 52. 60. 80. 45. 127. 112. 75. 95.   7. 170.
  94. 41. 60. 98. 80. 98. 85. 196. 136. 98.   7. 139. 93. 131.
 134. 131. 115. 125. 74. 161. 120. 125. 117. 131. 130. 200. 121. 110.
 142. 133. 126. 86. 127. 229. 87. 84. 94. 160. 80. 116. 98. 137.
 125. 171. 112. 101. 123. 70. 100. 93. 82. 90. 119. 194. 96. 155.
  80. 60.   3. 60. 89. 60. 60. 81. 60. 169. 94. 116. 95.
  90. 94. 60. 90. 102. 30.   6. 94. 85. 124. 13. 92. 86. 15.
  79. 87. 137. 20. 87. 84. 60. 102. 106. 114. 60. 107. 135. 54.
 103. 100. 60. 86. 60. 60. 40. 115. 60. 65. 86. 95. 119. 60.
  97. 105. 120. 60. 93. 60. 60. 60. 105. 60. 99. 83. 60. 263.
  60. 112. 60. 60. 60. 60. 112. 60. 13. 60. 99. 201. 60. 270.
  43. 87. 60. 60. 120.]

图 3-14 电影时长

电影评分= [9.6 9.5 9.5 9.4 9.4 9.4 9.4 9.4 9.4 9.3 9.3 9.3 9.3 9.3 9.3 9.3 9.3 9.3
 9.2 9.2 9.2 9.2 9.2 9.2 9.2 9.2 9.2 9.2 9.2 9.2 9.2 9.2 9.2 9.2 9.2 9.2
 9.2 9.2 9.2 9.2 9.2 7.4 8.5 9.2 9.2 9.2 8. 9.1 9.1 5.6 9.1 9.1 8.6 9.1
 5.7 9.1 9. 7.8 7.7 8.3 8. 8.3 4.4 8.1 9.1 9.1 9.1 9.1 9.1 6.4 6.8 6.8
 7.2 7. 9.1 9.1 6.8 6.8 6.9 6.9 7. 9.1 6.7 7. 6.6 6.8 6.9 9.1 7. 9.1
 9.1 5.5 5.8 9.1 5.1 5.3 9.1 9.1 9.1 9.1 7.4 7.4 9.1 7.5 7. 7.2 7.7 6.9
 5.6 9.1 9.1 9.1 9.1 7.3 9.1 8.3 9.1 8.3 8.3 5.7 5.7 6.5 9.1 7.2 8.1 5.1
 4.6 9.1 7.8 8.2 6.5 7.4 8.3 5.4 3.3 7.1 7.1 6.5 9.1 7.9 7.1 4.5 6.5 7.6
 6.1 5.3 8.2 8.2 7.6 6.3 7.4 6.3 6.6 8.8 6.2 7.1 8.2 8.9 6.3 8.2 7.9 5.3
 2.6 5.2 4. 6.8 4.6 5.8 7.9 7.4 7.2 8.6 7.6 7.7 8.4 5.6 6.3 6.2 8.4 9.6
 5.7 6.6 5.3 5.7 6.6 7.3 5.7 6.1 8.6 7.5 5.4 7.7 6.7 9.1 8. 6.2 8.6 6.
 6.4 4.7 9.1]

图 3-15 电影评分

4）使用 dict() 函数创建一个字典数据:dic_data2。

```
dic_data2 = dict(
    # fname 表示读取的文件
    fname = "电影数据.csv",
```

```
# delimiter 表示数据分隔符
delimiter = ',',
# dtype 表示输出的源数据类型
dtype = str,
# usecols 表示读取的列坐标
usecols = (1),
# skiprows 跳过前 x 行, 1 表示跳过第一行表头
skiprows = 1,
# encoding = 'utf-8'表示使用指定的字符集打开文件
encoding = 'utf-8',
# unpack 表示如果为 True, 读入属性将分别写入不同数组变量, False 读入数据只写入一
个数组变量, 默认为 False
unpack = True
)
```

5）使用 np.loadtxt 方法读取字典 dic_data2，并将指定列的数据统一赋值给 b，组成一个新的 ndarray 数组，如图 3-16 所示。

```
b = np.loadtxt(**dic_data2)
# 通过 a 的数组下标分别获取指定的字段数据：电影名称
print("电影名称=",b)
```

电影名称= ['肖申克的救赎' '控方证人' '美丽人生' '阿甘正传' '霸王别姬' '泰坦尼克号' '辛德勒的名单'
'新世纪福音战士劇場版: Air/真心为你 新世紀エヴァンゲリヲン劇場版 Ai' '银魂完结篇: 直到永远的万事屋 劇場版 銀魂 完結篇 万事屋よ'
'这个杀手不太冷' '灿烂人生' '疯狂动物城' '福音战士新劇場版: 破 エヴァンゲリヲン新劇場版:' '海豚湾']

图 3-16 电影名称

6）使用 np.max、np.min 和 np.average 方法计算最高投票数量、最低投票数量、最长电影时长、最短电影时长、最高评分、最低评分，如图 3-17 所示。

```
a0_maxVote = np.max(a[0])
a0_minVote = np.min(a[0])
a1_maxLength = np.max(a[1])
a1_minLength = np.min(a[1])
a2_maxScore = np.max(a[2])
a2_minScore = np.min(a[2])
a2_avgScore = np.average(a[2])
print("最高投票数量=",a0_maxVote)
print("最低投票数量=",a0_minVote)
print("最长电影时长=",a1_maxLength)
print("最短电影时长=",a1_minLength)
print("最高评分=",a2_maxScore)
print("最低评分=",a2_minScore)
```

最高投票数量= 692795.0
最低投票数量= 28.0
最长电影时长= 366.0
最短电影时长= 3.0
最高评分= 9.6
最低评分= 2.6

图 3-17 电影部分数据统计

7）获取投票数最多和最少的电影，如图 3-18 所示。

```
a0_max_lst = []
a0_min_lst = []
for i in range(len(a[0])):
    if (a[0][i] == 692795.0):
        a0_max_lst.append(b[i])
    elif (a[0][i] == 28.0):
        a0_min_lst.append(b[i])
print("最高投票数量的电影名称：",a0_max_lst)
print("最低投票数量的电影名称：",a0_min_lst)
```

最高投票数量的电影名称：['肖申克的救赎']
最低投票数量的电影名称：['13平']

图 3-18　最高、最低投票数量的电影名称

8）获取最长、最短电影时长的电影名称，如图 3-19 所示。

```
a1_max_lst = []
a1_min_lst = []
for i in range(len(a[1])):
    if (a[1][i] == 366.0):
        a1_max_lst.append(b[i])
    elif (a[1][i] == 3.0):
        a1_min_lst.append(b[i])
print("最长电影时长的电影名称：",a1_max_lst)
print("最短电影时长的电影名称：",a1_min_lst)
```

最长电影时长的电影名称：['灿烂人生 ']
最短电影时长的电影名称：['无翼鸟']

图 3-19　最长、最短电影时长的电影名称

9）获取评分最高和最低的电影名称，如图 3-20 所示。

```
a2_max_lst = []
a2_min_lst = []
for i in range(len(a[2])):
    if (a[2][i] == 9.6):
        a2_max_lst.append(b[i])
    elif (a[2][i] == 2.6):
        a2_min_lst.append(b[i])
print("最高评分电影名称：",a2_max_lst)
print("最低评分电影名称：",a2_min_lst)
```

最高评分电影名称：['肖申克的救赎', '1991年莫斯科摇']
最低评分电影名称：['18爱不爱之怦然']

图 3-20　最高、最低评分电影名称

练习题

（1）创建一个各元素的值为 0~16 的 3×3 矩阵。
（2）创建一个值为 20~69 的数组，步长为 2。
（3）生成一个 6×6 的对角矩阵。
（4）创建一个 5×10 的随机值数组，并找到最大值、最小值、平均值和标准差。
（5）创建一个四边为 1，中间为 0 的二维数组。
（6）创建一个四边为 0，中间为 1 的二维数组。
（7）如何计算 $((A+B)\times(-A/2))$？
（8）创建一个长度为 5 的数组，并做排序操作。
（9）求数组 a 的累计和和累计积。
（10）创建随机数组，并保存该数组为 NumPy 二进制文件，读取 NumPy 二进制文件。

项目 4 Pandas 数据分析和处理

【项目分析】

本项目旨在理解和掌握 Pandas 数据分析工具的基本概念和基础语法。具体内容如下。
1) Series（一维数据）与 DataFrame（二维数据）和内部结构及其实例。
2) 数据查询常用函数和基本属性及其实例。
3) 数据子集函数 at、iat、loc 和 iloc 及其实例。
4) 删除、填充和转换缺失值和空值的函数及其实例。
5) 连接与合并 Series 和 DataFrame 对象的函数及其实例。
6) 数据分类分组操作及其实例。
7) 多层索引和轴向转换实现数据重塑的函数及其实例。
8) 矩阵格式的数据按照行或者列进行组合的函数及其实例。
9) 时间数据的类和函数及其实例。
10) 数字分类的类和函数及其实例。
11) 读写 Excel 文件和 csv 格式文件的函数及其实例。
12) Pandas 数据分析案例介绍。

【知识准备】

4.1 Pandas 的基本概念

4.1.1 Pandas 基础理论和引用方法

Pandas 是建立在 NumPy 之上的能够支持多种语言的开源数据分析工具，是 Python 的重要数据分析库。Pandas 提供了强大、灵活、准确的数据结构 Series（一维数据）与 DataFrame（二维数据），能够高效地处理表格数据、时间序列数据和矩阵数据等同构或异构数据。与前面 NumPy 的引用方式类似，这里通过引用 Anaconda 自带的 Python 解释器就可以导入 Pandas 包，如图 4-1 所示。

图 4-1 导入 Pandas 包

4.1.2 Pandas 基本数据结构

（1）Series（一维数据）

Series 对象与 NumPy 的 ndarray 数组对象和 Python 的 list 列表对象类似，是一维数据结构。Series 可以保存多种不同的数据类型，并且拥有丰富的内置函数和属性。Series 结构包含 index、name 及其 values。Series 的 index 具有唯一性，index 可以是数字和字符，系统会自动将它们转化为一个类型 object，如图 4-2 所示。

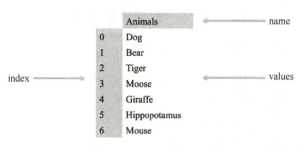

图 4-2　Series 结构示例

（2）DataFrame（二维数据）

DataFrame 是一个表格型的数据结构，包含一组有序的列，每列可以是不同的值类型（数值、字符串、布尔型等），DataFrame 既有行索引也有列索引，可以被看作是由 Series 组成的字典，如图 4-3 所示。

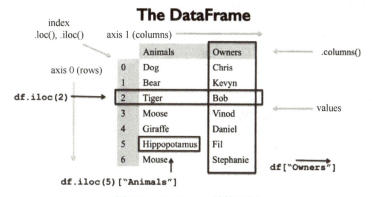

图 4-3　DataFrame 结构示例

4.2 Pandas 的基本用法

4.2.1 创建 Pandas 对象

Pandas 主要有两种数据类型对象：Series（一维数据）与 DataFrame（二维数据）。其中，DataFrame 可以包含一个或多个 Series。

【实例 4-1】使用 Pandas 的 Series 类创建 Series 对象，输出 Series 对象的值和属性并修改 Series 对象的索引。

```python
# 从 Pandas 包里面导入 Series
from Pandas import Series
# 创建一个字典数据，包含 3 个键：'Province','City'和'Code'及其对应值
data = {'Province':['Hebei', 'Henan', 'Sichuan'],
        'City':['Shijiazhuang', 'Zhengzhou', 'Chengdu'],
        'Code':[123, 456, 789]
        }
# 创建 Series 对象
# 使用 Series 类接收 data['Province']作为参数创建 Series 对象 s1
s1 = Series(data['Province'])
# 输出 Series 对象 s1 的索引、值和数据类型
print(s1)
print('输出 Series 对象 s1 的索引、值和数据类型')
# 输出 Series 对象 s1 的值
print(s1.values)
print('输出 Series 对象 s1 的值')
# 输出 Series 对象 s1 的索引
print(s1.index)
print('输出 Series 对象 s1 的索引')
# 为 Series 指定 index 为['A', 'B', 'C']
s1 = Series(data['Province'], index=['A', 'B', 'C'])
# 输出修改索引后的结果
print(s1)
print('''为 Series 指定 index 为['A', 'B', 'C']''')
```

输出如下：

```
0    Hebei
1    Henan
2    Sichuan
dtype: object
输出 Series 对象 s1 的索引、值和数据类型
['Hebei' 'Henan' 'Sichuan']
输出 Series 对象 s1 的值
RangeIndex(start=0, stop=3, step=1)
输出 Series 对象 s1 的索引
A    Hebei
B    Henan
C    Sichuan
dtype: object
为 Series 指定 index 为['A', 'B', 'C']
```

DataFrame 可以通过二维数组或者字典数据创建。其中，字典中的 value 值只能是一维数组或单个的简单数据类型。如果是数组，要求所有数组长度一致。如果是单个数据，则会使每行添加相同数据。下面将以字典数据创建 DataFrame 对象。

【实例 4-2】使用 Pandas 的 DataFrame 类创建 DataFrame 对象，输出 DataFrame 对象成员的值和类型，修改 DataFrame 对象的索引并实现矩阵的转置。

```python
import Pandas as pd
```

```python
# 从 Pandas 包里面导入 DataFrame
from Pandas import DataFrame
# 创建一个字典数据，包含 3 个键：'Province','City'和'Code'及其对应值。
data = {'Province':['Hebei', 'Henan', 'Sichuan'],
        'City':['Shijiazhuang', 'Zhengzhou', 'Chengdu'],
        'Code':[123, 456, 789]
        }
# 创建# DataFrame 对象
# 使用# DataFrame 类接收 data 作为参数创建 DataFrame 对象 df1
df1 = DataFrame(data)
# 输出 DataFrame 对象 df1 的索引、值和列名
print(df1)
print('输出 DataFrame 对象 df1 的索引、值和列名')
# 输出 Series 对象 df1['City']的索引、值和数据类型
print(df1['City'])
print('''输出 Series 对象 df1['City']的索引、值和数据类型''')
# 返回一个生成器用于迭代操作<generator object DataFrame.iterrows at 0x000001A9DC535580>
print(df1.iterrows())
print('返回一个生成器用于迭代操作<generator object DataFrame.iterrows at 0x000001A9DC535580>')
for row in df1.iterrows():
    # 输出 DataFrame 迭代器每一行数据的值
    print(row)
    # 输出 DataFrame 迭代器每一行数据第 1 列和第 2 列的值
    print(row[0], row[1])
    # 输出 DataFrame 迭代器每一行数据第 1 列和第 2 列的类型
    print(type(row[0]), type(row[1]))
    break
print('输出 DataFrame 迭代器每一行数据的值')
print('输出 DataFrame 迭代器每一行数据第 1 列和第 2 列的值')
print('输出 DataFrame 迭代器每一行数据第 1 列和第 2 列的类型')
# 构建几个 Series
s1 = pd.Series(data['Province'])
s2 = pd.Series(data['City'])
s3 = pd.Series(data['Code'])
# 为 DataFrame 指定 index 为['Province', 'City', 'Code']
df_new = pd.DataFrame([s1, s2, s3], index=['Province', 'City', 'Code'])
# 输出修改后的索引
print(df_new)
print('''为 DataFrame 指定 index 为['Province', 'City', 'Code']''')
# 将 DataFrame 对象 df_new 进行转置后输出
print(df_new.T)
print('将 DataFrame 对象 df_new 进行转置后输出')
```

输出如下：

```
   Province          City  Code
0     Hebei  Shijiazhuang   123
1     Henan     Zhengzhou   456
2   Sichuan       Chengdu   789
输出 DataFrame 对象 df1 的索引、值和列名
0    Shijiazhuang
```

```
1        Zhengzhou
2        Chengdu
Name: City, dtype: object
```
输出 Series 对象 df1['City']的索引、值和数据类型
`<generator object DataFrame.iterrows at 0x00000204D8D85580>`
返回一个生成器用于迭代操作`<generator object DataFrame.iterrows at 0x000001A9DC535580>`
```
(0, Province         Hebei
City        Shijiazhuang
Code                 123
Name: 0, dtype: object)
0 Province         Hebei
City        Shijiazhuang
Code                 123
Name: 0, dtype: object
```
`<class 'int'> <class 'pandas.core.Series.Series'>`
输出 DataFrame 迭代器每一行数据的值
输出 DataFrame 迭代器每一行数据第 1 列和第 2 列的值
输出 DataFrame 迭代器每一行数据第 1 列和第 2 列的类型
```
                    0          1          2
Province        Hebei      Henan    Sichuan
City     Shijiazhuang  Zhengzhou    Chengdu
Code              123        456        789
```
为 DataFrame 指定 index 为['Province', 'City', 'Code']
```
   Province          City  Code
0     Hebei  Shijiazhuang   123
1     Henan     Zhengzhou   456
2   Sichuan       Chengdu   789
```
将 DataFrame 对象 df_new 进行转置后输出

4.2.2 查看 Pandas 基本数据

Pandas 针对 Series 对象和 DataFrame 对象提供了多个可以查看其内部数据信息的常用函数,例如 head、tail 和 describe,以及查看基本属性的 index、ndim 和 info 等信息,如表 4-1 所示。

表 4-1 查看 Pandas 基本数据

函数或属性名称	描 述
head	查看前面行的数据
tail	查看后面行的数据
describe	查看数据统计摘要
index	查看数据索引
ndim	查看数据维度
info	查看数据基本信息

【实例 4-3】使用 Series 的 head、tail 和 describe 函数和 index、ndim 属性查看 Series 对象的基本信息。

```
import numpy as np
import pandas as pd
# 使用 pd.Series 创建 Series 对象 s1
```

```
s1 = pd.Series([1, 3, 5, np.nan, 6, 8])
# 查看 Series 对象 s1 的信息
print(s1)
print('查看 Series 对象 s1 的信息')
# 查看 Series 对象 s1 前 3 行的信息
print(s1.head(3))
print('查看 Series 对象 s1 前 3 行的信息')
# 查看 Series 对象 s1 后 3 行的信息
print(s1.tail(3))
print('查看 Series 对象 s1 后 3 行的信息')
# 查看 Series 对象 s1 的索引信息
print(s1.index)
print('查看 Series 对象 s1 的索引信息')
# 查看 Series 对象 s1 的维度信息
print(s1.ndim)
print('查看 Series 对象 s1 的维度信息')
# 查看 Series 对象 s1 的统计摘要
print(s1.describe())
print('查看 Series 对象 s1 的统计摘要')
```

输出如下：

```
0    1.0
1    3.0
2    5.0
3    NaN
4    6.0
5    8.0
dtype: float64
查看 Series 对象 s1 的信息
0    1.0
1    3.0
2    5.0
dtype: float64
查看 Series 对象 s1 前 3 行的信息
3    NaN
4    6.0
5    8.0
dtype: float64
查看 Series 对象 s1 后 3 行的信息
RangeIndex(start=0, stop=6, step=1)
查看 Series 对象 s1 的索引信息
1
查看 Series 对象 s1 的维度信息
count    5.000000
mean     4.600000
std      2.701851
min      1.000000
25%      3.000000
50%      5.000000
75%      6.000000
max      8.000000
dtype: float64
```

查看 Series 对象 s1 的统计摘要

【实例 4-4】 使用 DataFrame 的 head、tail 和 describe 函数和 index、ndim 和 info 属性查看 DataFrame 对象的基本信息。

```python
import numpy as np
import pandas as pd
# 使用 pd.DataFrame 类以字典数据作为参数创建 DataFrame 对象 df1
# 使用 pd.Timestamp 创建时间格式数据
# 使用 pd.Series 创建 Series 对象，其中 index=list(range(4)) 表示创建 4 个索引号，数据类型为 float32
# 使用 np.array 创建 4 个元素的数组，数据类型为 int32
# 使用 pd.Categorical 类创建 Categorical 类别对象，后期可以用于分类统计元素类别
df1 = pd.DataFrame({'A': 1.,
                    'B': pd.Timestamp('20210131'),
                    'C': pd.Series(2, index=list(range(4)), dtype='float32'),
                    'D': np.array([6] * 4, dtype='int32'),
                    'E': pd.Categorical(["Tom", "Simon", "John", "Joe"]),
                    'F': 'foo'})

# 查看 DataFrame 对象 df1 的信息
print(df1)
print('查看 DataFrame 对象 df1 的信息')
# 查看 DataFrame 对象 df1 前 3 行的信息
print(df1.head(3))
print('查看 DataFrame 对象 df1 前 3 行的信息')
# 查看 DataFrame 对象 df1 后 3 行的信息
print(df1.tail(3))
print('查看 DataFrame 对象 df1 后 3 行的信息')
# 查看 DataFrame 对象 df1 的索引信息
print(df1.index)
print('查看 DataFrame 对象 df1 的索引信息')
# 查看 DataFrame 对象 df1 的维度信息
print(df1.ndim)
print('查看 DataFrame 对象 df1 的维度信息')
# 查看 DataFrame 对象 df1 的基本信息
print(df1.info)
print('查看 DataFrame 对象 df1 的基本信息')
# 查看 DataFrame 对象 df1 的统计摘要
print(df1.describe())
print('查看 DataFrame 对象 df1 的统计摘要')
```

输出如下：

```
     A          B    C  D      E    F
0  1.0 2021-01-31  2.0  6    Tom  foo
1  1.0 2021-01-31  2.0  6  Simon  foo
2  1.0 2021-01-31  2.0  6   John  foo
3  1.0 2021-01-31  2.0  6    Joe  foo
查看 DataFrame 对象 df1 的信息
     A          B    C  D      E    F
0  1.0 2021-01-31  2.0  6    Tom  foo
1  1.0 2021-01-31  2.0  6  Simon  foo
```

```
2  1.0 2021-01-31  2.0  6   John  foo
查看 DataFrame 对象 df1 前 3 行的信息
     A          B    C    D   E     F
1  1.0 2021-01-31  2.0  6  Simon  foo
2  1.0 2021-01-31  2.0  6   John  foo
3  1.0 2021-01-31  2.0  6    Joe  foo
查看 DataFrame 对象 df1 后 3 行的信息
Int64Index([0, 1, 2, 3], dtype='int64')
查看 DataFrame 对象 df1 的索引信息
2
查看 DataFrame 对象 df1 的维度信息
<bound method DataFrame.info of
     A          B    C    D   E     F
0  1.0 2021-01-31  2.0  6    Tom  foo
1  1.0 2021-01-31  2.0  6  Simon  foo
2  1.0 2021-01-31  2.0  6   John  foo
3  1.0 2021-01-31  2.0  6    Joe  foo>
查看 DataFrame 对象 df1 的基本信息
         A    C    D
count  4.0  4.0  4.0
mean   1.0  2.0  6.0
std    0.0  0.0  0.0
min    1.0  2.0  6.0
25%    1.0  2.0  6.0
50%    1.0  2.0  6.0
75%    1.0  2.0  6.0
max    1.0  2.0  6.0
查看 DataFrame 对象 df1 的统计摘要
```

4.2.3　Pandas 索引和切片

Pandas 拥有多种索引方式，主要有 at、iat、loc 和 iloc。这些 Pandas 数据访问方法通过切块、切片和索引的形式可以高效、准确地获得 Pandas 对象的子集，见表 4-2。

表 4-2　索引和切片

函 数 名 称	描　　述
at	通过标签来定位某个值
iat	通过数组下标来定位某个值
loc	通过标签标定位某个切块
iloc	通过数组下标定位某个切块

【实例 4-5】使用 Pandas 的 at 和 iat 函数实现 DataFrame 对象的索引和赋值。

```
import numpy as np
import pandas as pd
# 使用 pd.date_range 函数创建时间格式数据，periods 表示创建的个数
dates = pd.date_range('1/10/2021', periods=8)
print(dates)
print('使用 pd.date_range 函数创建时间格式数据，periods 表示创建的个数')
# 使用 pd.DataFrame 创建 DataFrame 对象 df
```

```python
# np.random.randn(8, 4)表示创建一个 8 行 4 列的随机矩阵
# index 和 columns 分别表示 DataFrame 对象 df 的行和列索引
df = pd.DataFrame(np.random.randn(8, 4),
                  index=dates, columns=['A', 'B', 'C', 'D'])
print(df)
print('使用 pd.DataFrame 创建 DataFrame 对象 df')
print('np.random.randn(8, 4)表示创建一个 8 行 4 列的随机矩阵')
print('index 和 columns 分别表示 DataFrame 对象 df 的行和列索引')
# 使用 DataFrame 的 at 函数实现 2021-01-10 行和 A 列的索引
print(df.at['2021-01-10','A'])
print('使用 DataFrame 的 at 函数实现 2021-01-10 行和 A 列的索引')
# 使用 DataFrame 的 iat 函数实现第一行第一列的索引
print(df.iat[0,0])
print('使用 DataFrame 的 iat 函数实现第一行第一列的索引')
# 使用 DataFrame 的 at 函数实现 2021-01-10 行和 A 列的赋值
df.at['2021-01-10','A']=888
print(df)
print('使用 DataFrame 的 at 函数实现 2021-01-10 行和 A 列的赋值')
```

输出如下：

```
DatetimeIndex(['2021-01-10', '2021-01-11', '2021-01-12', '2021-01-13',
'2021-01-14', '2021-01-15', '2021-01-16', '2021-01-17'], dtype='datetime64
[ns]', freq='D')
使用 pd.date_range 函数创建时间格式数据，periods 表示创建的个数
                   A         B         C         D
2021-01-10  0.529554  0.699422  1.138653 -1.037976
2021-01-11  0.008222  0.154153 -1.054278 -0.871804
2021-01-12 -0.769207  0.477511 -0.472494  1.148657
2021-01-13  0.814006 -1.093083  0.763773 -0.993329
2021-01-14  1.492796  0.469311 -0.152022  1.157729
2021-01-15 -0.248286 -0.107424  0.246081 -0.639680
2021-01-16  1.128357 -0.038286  1.694647  0.473954
2021-01-17 -0.006426  0.094023  0.696887  0.783493
使用 pd.DataFrame 创建 DataFrame 对象 df
np.random.randn(8, 4)表示创建一个 8 行 4 列的随机矩阵
index 和 columns 分别表示 DataFrame 对象 df 的行和列索引
0.5295543429109918
使用 DataFrame 的 at 函数实现 2021-01-10 行和 A 列的索引
0.5295543429109918
使用 DataFrame 的 iat 函数实现第一行第一列的索引
                     A         B         C         D
2021-01-10  888.000000  0.699422  1.138653 -1.037976
2021-01-11    0.008222  0.154153 -1.054278 -0.871804
2021-01-12   -0.769207  0.477511 -0.472494  1.148657
2021-01-13    0.814006 -1.093083  0.763773 -0.993329
2021-01-14    1.492796  0.469311 -0.152022  1.157729
2021-01-15   -0.248286 -0.107424  0.246081 -0.639680
2021-01-16    1.128357 -0.038286  1.694647  0.473954
2021-01-17   -0.006426  0.094023  0.696887  0.783493
使用 DataFrame 的 at 函数实现 2021-01-10 行和 A 列的赋值
```

【**实例 4-6**】 使用 Pandas 的 loc 和 iloc 函数实现 DataFrame 对象的索引和赋值。

```python
import numpy as np
import pandas as pd
# 使用 pd.date_range 函数创建时间格式数据，periods 表示创建的个数
dates = pd.date_range('1/10/2021', periods=8)
print(dates)
print('使用 pd.date_range 函数创建时间格式数据，periods 表示创建的个数')
# 使用 pd.DataFrame 创建 DataFrame 对象 df
# np.random.randn(8, 4) 表示创建一个 8 行 4 列的随机矩阵
# index 和 columns 分别表示 DataFrame 对象 df 的行和列索引
df = pd.DataFrame(np.random.randn(8, 4),
                  index=dates, columns=['A', 'B', 'C', 'D'])
print(df)
print('使用 pd.DataFrame 创建 DataFrame 对象 df')
print('np.random.randn(8, 4) 表示创建一个 8 行 4 列的随机矩阵')
print('index 和 columns 分别表示 DataFrame 对象 df 的行和列索引')
# 使用 DataFrame 的 loc 函数实现 2021-01-10 行和 A 列的索引
print(df.loc['2021-01-10','A'])
print('使用 DataFrame 的 loc 函数实现 2021-01-10 行和 A 列的索引')
# 使用 DataFrame 的 loc 函数实现 B 列的索引
print(df.loc[:,'B'])
print('使用 DataFrame 的 loc 函数实现 B 列的索引')
# 使用 DataFrame 的 loc 函数实现 2021-01-10 行的索引
print(df.loc['2021-01-10',:])
print('使用 DataFrame 的 loc 函数实现 2021-01-10 行的索引')
# 使用 DataFrame 的 iloc 函数实现第一行第一列的索引
print(df.iloc[0,0])
print('使用 DataFrame 的 iloc 函数实现第一行第一列的索引')
# 使用 DataFrame 的 at 函数实现所有行和 C 列的赋值
df.loc[:,'C']=999
print(df)
print('使用 DataFrame 的 at 函数实现所有行和 C 列的赋值')
```

输出如下：

```
DatetimeIndex(['2021-01-10', '2021-01-11', '2021-01-12', '2021-01-13',
'2021-01-14', '2021-01-15', '2021-01-16', '2021-01-17'], dtype='datetime64
[ns]', freq='D')
使用 pd.date_range 函数创建时间格式数据，periods 表示创建的个数
                   A         B         C         D
2021-01-10  -0.211788 -0.734653  0.175871  0.665371
2021-01-11   0.251415 -0.050060  1.158070  0.482428
2021-01-12   1.110582  0.434448  1.032795 -0.145543
2021-01-13   0.269975  1.468902 -0.818383  1.020840
2021-01-14   0.035391 -1.015809 -0.439783  0.126292
2021-01-15   0.296506 -0.200003 -0.859036  1.496871
2021-01-16   1.071159 -0.382932  1.701281 -1.074745
2021-01-17  -0.608950  0.118984  0.414718  1.684679
使用 pd.DataFrame 创建 DataFrame 对象 df
np.random.randn(8, 4) 表示创建一个 8 行 4 列的随机矩阵
index 和 columns 分别表示 DataFrame 对象 df 的行和列索引
-0.21178805975802115
使用 DataFrame 的 loc 函数实现 2021-01-10 行和 A 列的索引
2021-01-10   -0.734653
```

```
2021-01-11   -0.050060
2021-01-12    0.434448
2021-01-13    1.468902
2021-01-14   -1.015809
2021-01-15   -0.200003
2021-01-16   -0.382932
2021-01-17    0.118984
Freq: D, Name: B, dtype: float64
```
使用 DataFrame 的 loc 函数实现 B 列的索引
```
A   -0.211788
B   -0.734653
C    0.175871
D    0.665371
Name: 2021-01-10 00:00:00, dtype: float64
```
使用 DataFrame 的 loc 函数实现 2021-01-10 行的索引
-0.21178805975802115
使用 DataFrame 的 iloc 函数实现第一行第一列的索引
```
                   A          B        C         D
2021-01-10   -0.211788  -0.734653   999    0.665371
2021-01-11    0.251415  -0.050060   999    0.482428
2021-01-12    1.110582   0.434448   999   -0.145543
2021-01-13    0.269975   1.468902   999    1.020840
2021-01-14    0.035391  -1.015809   999    0.126292
2021-01-15    0.296506  -0.200003   999    1.496871
2021-01-16    1.071159  -0.382932   999   -1.074745
2021-01-17   -0.608950   0.118984   999    1.684679
```
使用 DataFrame 的 at 函数实现所有行和 C 列的赋值

4.2.4　Pandas 缺失值和空值处理

Pandas 提供了可以删除、填充和转换缺失值和空值的函数，分别是 dropna、fillna、isna 和 isnull，见表 4-3。在 DataFrame 中使用 nan 或 NaT（缺失时间）表示缺失值，在 Series 中使用 none 和 nan 表示缺失值，在 Pandas 中的空值是""。

表 4-3　Pandas 缺失值和空值处理

函 数 名 称	描　　述
dropna	删除缺失值
fillna	填充缺失值
isna	判断缺失值
isnull	判断空值

【实例 4-7】使用 Pandas 的 dropna、fillna、isna 和 isnull 函数删除、填充和转换缺失值或空值，如下所示。

```
import numpy as np
import pandas as pd
# 使用 pd.DataFrame 创建 DataFrame 对象 df
# 使用 np.nan 表示缺失值
# 使用 pd.NaT 表示缺失时间值
```

```python
df = pd.DataFrame({"name": ['Alex', 'Bill', 'Cute'],
            "tool": [np.nan, 'bike', 'car'],
             "time": [pd.NaT, pd.Timestamp("2021-02-10"),pd.NaT]})
print(df)
print('使用 pd.DataFrame 创建 DataFrame 对象 df')
print('使用 np.nan 表示缺失值')
print('使用 pd.NaT 表示缺失时间值')
# 使用 dropna 函数删除有缺失值的行
print(df.dropna())
print('使用 dropna 函数删除有缺失值的行')
# 使用 fillna 函数以 0 填充缺失值
print(df.fillna(0))
print('使用 fillna 函数以 0 填充缺失值')
# 使用 isna 函数判断是否存在缺失值，并以布尔值作为掩码表示
print(df.isna())
print('使用 isna 函数判断是否存在缺失值，并以布尔值作为掩码表示')
# 使用 isnull 函数判断是否存在空值或缺失值，并以布尔值作为掩码表示
print(df.isnull())
print('使用 isnull 函数判断是否存在空值或缺失值，并以布尔值作为掩码表示')
```

输出如下：

```
   name tool       time
0  Alex  NaN        NaT
1  Bill bike 2021-02-10
2  Cute  car        NaT
使用 pd.DataFrame 创建 DataFrame 对象 df
使用 np.nan 表示缺失值
使用 pd.NaT 表示缺失时间值
   name tool       time
1  Bill bike 2021-02-10
使用 dropna 函数删除有缺失值的行
   name tool                time
0  Alex    0                   0
1  Bill bike 2021-02-10 00:00:00
2  Cute  car                   0
使用 fillna 函数以 0 填充缺失值
    name   tool   time
0  False   True   True
1  False  False  False
2  False  False   True
使用 isna 函数判断是否存在缺失值，并以布尔值作为掩码表示
    name   tool   time
0  False   True   True
1  False  False  False
2  False  False   True
使用 isnull 函数判断是否存在空值或缺失值，并以布尔值作为掩码表示
```

4.2.5 Pandas 连接和合并数据

Pandas 提供了可以连接和合并操作 Series、DataFrame 对象的函数，分别是 concat、join 和

append，见表 4-4。

表 4-4　Pandas 连接和合并数据

函 数 名 称	描　　述
concat	合并数据
join	连接数据（内连接、外连接）
append	尾部添加数据

【实例 4-8】 使用 Pandas 的 concat、join 和 append 函数连接与合并 Series 和 DataFrame 对象数据。

```
import numpy as np
import pandas as pd
df = pd.DataFrame(np.random.randn(8, 4))
print(df)
# 将 DataFrame 对象 df 的数据切分为 3 块，具体如下
blocks = [df[:2], df[2:6], df[6:]]
print(blocks[0])
print('将 DataFrame 对象 df 的数据切分为 3 块:blocks[0]')
print(blocks[1])
print('将 DataFrame 对象 df 的数据切分为 3 块:blocks[1]')
print(blocks[2])
print('将 DataFrame 对象 df 的数据切分为 3 块:blocks[2]')
# 使用 concat 函数将 blocks 合并一起
print(pd.concat(blocks))
print('使用 concat 函数将 blocks 合并在一起')
# 创建两个 DataFrame 对象 df_left 和 df_right 用于合并操作
df_left = pd.DataFrame({'key': ['hello', 'hello'], 'left_value': [1, 2]})
df_right = pd.DataFrame({'key':['hello', 'hello'], 'right_value': [4, 5]})
print(df_left)
print('创建 DataFrame 对象 df_left')
print(df_right)
print('创建 DataFrame 对象 df_right')
# 使用 merge 函数将 df_left 和 df_right 通过 key 进行合并
df_merged = pd.merge(df_left, df_right, on='key')
print(df_merged)
print('使用 merge 函数将 df_left 和 df_right 通过 key 进行合并')
# 使用 concat 函数以 join="inner"内连接的形式将 df_left, df_right 进行合并
df_join_inner = pd.concat([df_left, df_right], axis=1, join="inner")
print(df_join_inner)
print('使用 concat 函数以 join="inner"内连接的形式将 df_left, df_right 进行合并')
# 使用 concat 函数以 join="outer"外连接的形式将 df_left, df_right 进行合并
df_join_outer = pd.concat([df_left, df_right], axis=1, join="outer")
print(df_join_outer)
print('使用 concat 函数以 join="outer"外连接的形式将 df_left, df_right 进行合并')
# 创建 DataFrame 对象 df2
df2 = pd.DataFrame(np.random.randn(6, 4), columns=['A', 'B', 'C', 'D'])
print(df2)
print('创建 DataFrame 对象 df2')
# 使用 loc 函数切去第 2 行的数据
s = df2.iloc[1]
```

```
print(s)
print('使用 loc 函数切去第 2 行的数据')
# 使用 append 将 s 添加在尾部
df2_appended = df2.append(s, ignore_index=True)
print(df2_appended)
print('使用 append 将 s 添加在尾部')
```

输出如下：

```
          0         1         2         3
0 -1.300848 -0.439227 -0.798490 -0.502218
1 -1.139931  1.579842 -0.498711 -2.003730
2 -0.271485  0.450550 -1.261194 -1.093589
3 -0.713860  0.826725  1.025101 -1.745289
4 -0.738310  0.730229  1.045013 -0.740194
5  0.819079 -2.762642 -1.674347 -1.220915
6 -0.659197 -0.972530  1.455212 -0.361011
7 -0.347483 -0.464705 -0.130585 -0.602087
          0         1         2         3
0 -1.300848 -0.439227 -0.798490 -0.502218
1 -1.139931  1.579842 -0.498711 -2.003730
将 DataFrame 对象 df 的数据切分为 3 块:blocks[0]
          0         1         2         3
2 -0.271485  0.450550 -1.261194 -1.093589
3 -0.713860  0.826725  1.025101 -1.745289
4 -0.738310  0.730229  1.045013 -0.740194
5  0.819079 -2.762642 -1.674347 -1.220915
将 DataFrame 对象 df 的数据切分为 3 块:blocks[1]
          0         1         2         3
6 -0.659197 -0.972530  1.455212 -0.361011
7 -0.347483 -0.464705 -0.130585 -0.602087
将 DataFrame 对象 df 的数据切分为 3 块:blocks[2]
          0         1         2         3
0 -1.300848 -0.439227 -0.798490 -0.502218
1 -1.139931  1.579842 -0.498711 -2.003730
2 -0.271485  0.450550 -1.261194 -1.093589
3 -0.713860  0.826725  1.025101 -1.745289
4 -0.738310  0.730229  1.045013 -0.740194
5  0.819079 -2.762642 -1.674347 -1.220915
6 -0.659197 -0.972530  1.455212 -0.361011
7 -0.347483 -0.464705 -0.130585 -0.602087
使用 concat 函数将 blocks 合并在一起
     key  left_value
0  hello           1
1  hello           2
创建 DataFrame 对象 df_left
     key  right_value
0  hello            4
1  hello            5
创建 DataFrame 对象 df_right
     key  left_value  right_value
0  hello           1            4
1  hello           1            5
```

```
2  hello           2           4
3  hello           2           5
```
使用 merge 函数将 df_left 和 df_right 通过 key 进行合并
```
   key  left_value   key  right_value
0  hello          1  hello           4
1  hello          2  hello           5
```
使用 concat 函数以 join="inner" 内连接的形式将 df_left, df_right 进行合并
```
   key  left_value   key  right_value
0  hello          1  hello           4
1  hello          2  hello           5
```
使用 concat 函数以 join="outer" 外连接的形式将 df_left, df_right 进行合并
```
          A         B         C         D
0 -0.351426 -0.380249 -0.834499 -1.371192
1  0.879342 -0.246552  1.782550 -0.971888
2 -0.642381  1.245022  0.040049  1.398222
3  0.597150 -0.869845  0.369890 -0.796211
4  0.036746  0.401852  0.549792  0.678376
5  0.055160 -1.266145 -1.326845 -0.482364
```
创建 DataFrame 对象 df2
```
A    0.879342
B   -0.246552
C    1.782550
D   -0.971888
Name: 1, dtype: float64
```
使用 loc 函数切去第 2 行的数据
```
          A         B         C         D
0 -0.351426 -0.380249 -0.834499 -1.371192
1  0.879342 -0.246552  1.782550 -0.971888
2 -0.642381  1.245022  0.040049  1.398222
3  0.597150 -0.869845  0.369890 -0.796211
4  0.036746  0.401852  0.549792  0.678376
5  0.055160 -1.266145 -1.326845 -0.482364
6  0.879342 -0.246552  1.782550 -0.971888
```
使用 append 将 s 添加在尾部

4.2.6 Pandas 分组

Pandas 使用 "group by" 来对数据进行分类分组操作。

【实例 4-9】 使用 Pandas 的 group by 实现分组并求和。

```
import numpy as np
import pandas as pd
# 使用 DataFrame 创建 DataFrame 对象 df
df = pd.DataFrame({'A': ['hello', 'world', 'hello', 'world','hello', 'world', 'hello', 'hello'],
        'B':['one', 'one', 'two', 'three','two', 'two', 'one', 'three'],
        'C': np.random.randn(8),
        'D': np.random.randn(8)})
print(df)
print('使用 DataFrame 创建 DataFrame 对象 df')
```

```
# 使用 group by 根据 A 分组之后，再进行计算求和
df_sum = df.groupby('A').sum()
print(df_sum)
print('使用 group by 根据 A 分组之后，再进行计算求和')
# 使用 group by 分成 A,B 两组之后，再进行计算求和
df_sum2 = df.groupby(['A', 'B']).sum()
print(df_sum2)
print('使用 group by 分成 A,B 两组之后，再进行计算求和')
```

输出如下：

```
       A      B         C         D
0  hello    one -1.116940 -0.769008
1  world    one  0.113800  1.664659
2  hello    two -0.944012 -0.319312
3  world  three  0.156854 -1.231204
4  hello    two  0.042495  0.112051
5  world    two  0.661272 -0.431116
6  hello    one -0.945304  0.421378
7  hello  three  0.680863 -0.697473
使用 DataFrame 创建 DataFrame 对象 df
              C         D
A
hello -2.282897 -1.252364
world  0.931926  0.002339
使用 group by 根据 A 分组之后，再进行计算求和
                    C         D
A     B
hello one   -2.062243 -0.347630
      three  0.680863 -0.697473
      two   -0.901517 -0.207261
world one    0.113800  1.664659
      three  0.156854 -1.231204
      two    0.661272 -0.431116
使用 group by 分成 A,B 两组之后，再进行计算求和
```

4.2.7 Pandas 重塑

Pandas 提供了多层索引和轴向转换函数实现数据的重塑，分别是 stack、unstack 和 pivot，见表 4-5。

表 4-5 Pandas 重塑

函 数 名 称	描 述
stack	将列压缩至一层
unstack	stack 的逆向操作
pivot	按矩阵轴进行转向

【实例 4-10】 使用 Pandas 的 stack、unstack 和 pivot 函数实现多层索引和轴向转换的数据重塑。

```
import numpy as np
```

```python
import pandas as pd
# 使用 zip 创建元组数据,并使用 list 封装为列表数组
tuples = list(zip(*[['hello', 'hello', 'world', 'world',
                     'tom', 'tom', 'sam', 'sam'],
                    ['one', 'two', 'one', 'two',
                     'one', 'two', 'one', 'two']]))
print(tuples)
print('使用 zip 创建元组数据,并使用 list 封装为列表数组')
# 使用 MultiIndex 的 from_tuples 函数创建多行索引
index = pd.MultiIndex.from_tuples(tuples, names=['first', 'second'])
print(index)
print('使用 MultiIndex 的 from_tuples 函数创建多行索引')
# 使用 DataFrame 创建 8 行 2 列的 DataFrame 对象
df = pd.DataFrame(np.random.randn(8, 2), index=index, columns=['A', 'B'])
print(df)
print('使用 DataFrame 创建 8 行 2 列的 DataFrame 对象')
# 获取前 3 行的数据
df2 = df[:3]
print(df2)
print('获取前 3 行的数据')
# 使用 stack 函数将列压缩至一层
stacked = df2.stack()
print(stacked)
print('使用 stack 函数将列压缩至一层')
# 使用 unstack 函数做逆向操作
unstacked = stacked.unstack()
print(unstacked)
print('使用 unstack 函数做逆向操作')
# 创建一个 DataFrame 对象 df3
df3 = pd.DataFrame({'first': ['one','one','one','two','two','two'],
                    'second': ['A', 'B', 'C', 'A', 'B', 'C'],
                    'third': [1, 2, 3, 4, 5, 6]})
print(df3)
print('创建一个 DataFrame 对象 df3')
# 使用 pivot 函数实现轴转换
df_pivot = df3.pivot(index='first', columns='second', values='third')
print(df_pivot)
print('使用 pivot 函数实现轴转换')
```

输出如下:

```
[('hello', 'one'), ('hello', 'two'), ('world', 'one'), ('world', 'two'), ('tom', 'one'), ('tom', 'two'), ('sam', 'one'), ('sam', 'two')]
使用 zip 创建元组数据,并使用 list 封装为列表数组
MultiIndex([('hello', 'one'),
            ('hello', 'two'),
            ('world', 'one'),
            ('world', 'two'),
            (  'tom', 'one'),
            (  'tom', 'two'),
            (  'sam', 'one'),
            (  'sam', 'two')],
           names=['first', 'second'])
```

使用 MultiIndex 的 from_tuples 函数创建多行索引
```
                     A         B
first second
hello one    -1.077076  -0.269088
      two    -0.185296  -1.021114
world one    -0.253961  -0.229375
      two     0.730340   0.478113
tom   one     1.119313   0.227409
      two    -0.451558   1.550579
sam   one    -0.317717   0.920642
      two     0.217015  -0.066670
```
使用 DataFrame 创建 8 行 2 列的 DataFrame 对象
```
                     A         B
first second
hello one    -1.077076  -0.269088
      two    -0.185296  -1.021114
world one    -0.253961  -0.229375
```
获取前 3 行的数据
```
first  second
hello  one     A   -1.077076
               B   -0.269088
       two     A   -0.185296
               B   -1.021114
world  one     A   -0.253961
               B   -0.229375
dtype: float64
```
使用 stack 函数将列压缩至一层
```
                     A         B
first second
hello one    -1.077076  -0.269088
      two    -0.185296  -1.021114
world one    -0.253961  -0.229375
```
使用 unstack 函数做逆向操作
```
   first second third
0  one    A      1
1  one    B      2
2  one    C      3
3  two    A      4
4  two    B      5
5  two    C      6
```
创建一个 DataFrame 对象 df3
```
second  A  B  C
first
one     1  2  3
two     4  5  6
```
使用 pivot 函数实现轴转换

4.2.8 Pandas 数据透视表

Pandas 的数据透视表是数据分析的常用工具之一,用于汇聚一个或多个数据表格。通过将矩阵格式的数据按照行或者列进行组合,能够非常高效地实现数据汇总,主要函数是 pivot_table。

【实例 4-11】 使用 Pandas 的 pivot_table 函数创建透视表,如下所示。

```
import pandas as pd
    df = pd.DataFrame({'money': [18.99, 12.34, 23.01, 25.68, 26.59, 27.29],
                    'extra_money': [1.11, 1.65, 3.6, 3.41, 3.71, 4.81],
                    'matchbox': ['No', 'Yes', 'No', 'No', 'No', 'Yes'],
                    'day': ['Fri', 'Sat', 'Sun', 'Thur', 'Fri', 'Sun'],
                    'times': [3, 4, 4, 3, 5, 5],
                    'meals': ['dinner', 'dinner', 'dinner', 'dinner', 'dinner', 'dinner']
                    })
    # 使用 pivot_table 实现数据透视表
    # ['extra_money', 'times'] 表示使用 extra_money 和 times 进行聚合,
margins=True 表示部分总计(添加名为 All 的行和列标签,All 的值是均值)
    df_pivot1 = df.pivot_table(['extra_money', 'times'], index=['meals', 'day'], columns=['matchbox'],
                    margins=True)
    print(df_pivot1)
    print('使用 pivot_table 实现数据透视表')
    print('''['extra_money', 'times']表示使用 extra_money 和 times 进行聚合,
margins=True 表示部分总计(添加名为 All 的行和列标签,All 的值是均值)''')
```

输出如下:

```
            extra_money              times
matchbox    No      Yes     All     No      Yes     All
meals day
dinner Fri  2.4100  NaN     2.410000    4.00    NaN     4.0
       Sat  NaN     1.65    1.650000    NaN     4.0     4.0
       Sun  3.6000  4.81    4.205000    4.00    5.0     4.5
       Thur 3.4100  NaN     3.410000    3.00    NaN     3.0
All         2.9575  3.23    3.048333    3.75    4.5     4.0
```

使用 pivot_table 实现数据透视表
['extra_money', 'times']表示使用 extra_money 和 times 进行聚合,margins=True 表示部分总计(添加名为 All 的行和列标签,All 的值是均值)

4.2.9 Pandas 时间序列

Pandas 提供了专门用于处理时间数据的函数,主要包括处理固定时间跨度的序列、整合和转换时间的频率以及处理时间的增量,见表 4-6。

表 4-6 Pandas 时间序列

类 或 函 数	描 述
Timestamp、to_datetime	时间戳
DatetimeIndex、date_range	时间戳索引
Period	时间段
PeriodIndex、period_range	时间段索引

【实例 4-12】 使用 Pandas 的 Timestamp 和 Period 创建时间戳和时间段,并输出 DatetimeIndex

和 PeriodIndex 的索引值。

```
import pandas as pd
# 使用 Timestamp 类创建时间戳 1
ts1 = pd.Timestamp(2021,9,16)
print(ts1)
print('使用 Timestamp 类创建时间戳 1')
# 使用 Timestamp 类创建时间戳 2
ts2 = pd.Timestamp('2021-9-16')
print(ts2)
print('使用 Timestamp 类创建时间戳 2')
# 使用 Period 类创建时间间隔跨度
p1 = pd.Period("2021-09")
print(p1)
print('使用 Period 类创建时间间隔跨度')
p2 = pd.Period("2021-09", freq="D")
print(p2)
print('使用 Period 类创建时间间隔跨度,freq="D"表示精确到天')
#索引后会自动强制转为 DatetimeIndex 和 PeriodIndex。
dates = [pd.Timestamp("2021-09-01"), pd.Timestamp("2021-09-02"), pd.
Timestamp("2021-09-03"), pd.Timestamp("2021-09-04")]
ts3 = pd.Series(data=["Tom", "Bob", "Mary", "James"], index=dates)
print(ts3.index)
print('索引后会自动强制转为 DatetimeIndex')
p3 = [pd.Period("2021-09"), pd.Period("2021-10"), pd.Period("2021-11"),
pd.Period("2021-12")]
ts4 = pd.Series(data=["Tom", "Bob", "Mary", "James"], index=p3)
print(ts4.index)
print('索引后会自动强制转为 PeriodIndex')
```

输出如下:

```
2021-09-16 00:00:00
使用 Timestamp 类创建时间戳 1
2021-09-16 00:00:00
使用 Timestamp 类创建时间戳 2
2021-09
使用 Period 类创建时间间隔跨度
2021-09-01
使用 Period 类创建时间间隔跨度,freq="D"表示精确到天
DatetimeIndex(['2021-09-01', '2021-09-02', '2021-09-03', '2021-09-04'],
dtype='datetime64[ns]', freq=None)
索引后会自动强制转为 DatetimeIndex
PeriodIndex(['2021-09', '2021-10', '2021-11', '2021-12'], dtype='period[M]',
freq='M')
索引后会自动强制转为 PeriodIndex
```

【**实例 4-13**】 使用 Pandas 的 to_datetime 函数转换文本时间数据和 UNIX 时间数据。

```
import pandas as pd
# 使用 to_datetime 函数将文本时间数据转换为时间数据格式
dt1 = pd.to_datetime(pd.Series(["Jul 20, 2021", "2021-09-10", None]))
print(dt1)
print('使用 to_datetime 函数将文本时间数据转换为时间数据格式')
```

```python
# 使用 to_datetime 函数将文本时间数据转换为时间数据格式
dt2 = pd.to_datetime(["2021/11/23", "2021.12.30"])
print(dt2)
print('使用 to_datetime 函数将文本时间数据转换为时间数据格式')
# 使用 to_datetime 函数将 UNIX 时间转换为时间数据格式，unit="s"表示以秒为单位
dt3 = pd.to_datetime([1649726149, 1949206503, 1849392902], unit="s")
print(dt3)
print('使用 to_datetime 函数将 UNIX 时间转换为时间数据格式，unit="s"表示以秒为单位')
# 使用 to_datetime 函数将 UNIX 时间转换为时间数据格式，unit="ms"表示以毫秒为单位
dt4 = pd.to_datetime([1649450105908, 1749680105287, 1849120105363], unit="ms")
print(dt4)
print('使用 to_datetime 函数将 UNIX 时间转换为时间数据格式，unit="ms"表示以毫秒为单位')
```

输出如下：

```
0   2021-07-20
1   2021-09-10
2          NaT
dtype: datetime64[ns]
使用 to_datetime 函数将文本时间数据转换为时间数据格式
DatetimeIndex(['2021-11-23','2021-12-30'], dtype='datetime64[ns]', freq=None)
使用 to_datetime 函数将文本时间数据转换为时间数据格式
DatetimeIndex(['2022-04-12 01:15:49', '2031-10-08 06:15:03',
               '2028-08-09 00:15:02'],
              dtype='datetime64[ns]', freq=None)
使用 to_datetime 函数将 UNIX 时间转换为时间数据格式，unit="s"表示以秒为单位
DatetimeIndex(['2022-04-08 20:35:05.908000', '2025-06-11 22:15:05.287000',
               '2028-08-05 20:28:25.363000'],
              dtype='datetime64[ns]', freq=None)
使用 to_datetime 函数将 UNIX 时间转换为时间数据格式，unit="ms"表示以毫秒为单位
```

【实例 4-14】 使用 Pandas 的 date_range 和 period_range 函数创建时间戳跨度数据。

```python
import pandas as pd
# 使用 date_range 函数创建从 2021-9-20 开始，后 10 天的时间戳
dr1 = pd.date_range("2021-9-20", periods=10)
print(dr1)
print('使用 date_range 函数创建从 2021-9-20 开始，后 10 天的时间戳')
# 使用 date_range 函数创建从 2021-10-10 开始，后 10 天的时间戳
dr2 = pd.period_range("2021-10-10", periods=10)
print(dr2)
print('使用 date_range 函数创建从 2021-10-10 开始，后 10 天的时间戳')
```

输出如下：

```
DatetimeIndex(['2021-09-20', '2021-09-21', '2021-09-22', '2021-09-23',
'2021-09-24', '2021-09-25', '2021-09-26', '2021-09-27','2021-09-28', '2021-09-29'], dtype='datetime64[ns]', freq='D')
使用 date_range 函数创建从 2021-9-20 开始，后 10 天的时间戳
PeriodIndex(['2021-10-10', '2021-10-11', '2021-10-12', '2021-10-13',
'2021-10-14', '2021-10-15', '2021-10-16', '2021-10-17','2021-10-18', '2021-10-19'],dtype='period[D]', freq='D')
使用 date_range 函数创建从 2021-10-10 开始，后 10 天的时间戳
```

4.2.10 Pandas 分类

Pandas 使用 Categorical 类或者 astype('category')函数将数据进行分类,其目的是将类别信息转化成数值信息以方便统计。

【实例 4-15】 使用 Pandas 的 astype('category')函数和 Categorical 类实现 Categorical 类型转换,并分类统计数据。

```
import pandas as pd
# 使用 DataFrame 类创建 DataFrame 对象 df
df = pd.DataFrame({'name': ['Simon', 'Dyson', 'Kobe', 'Joy'],
                   'score': ['a', 'b', 'b', 'b'],
                   'gender': ['Female', 'Male', 'Male', 'Female']},
                  columns = ['name', 'score', 'gender'])
# 使用 astype 函数将 df 转换为 Categorical 类型
ct2 = df.astype('category')
print(ct2.dtypes)
print('使用 astype 函数将 df 转换为 Categorical 类型')
# 使用 Categorical 类将 df['score'] 转换为 Categorical 类型
ct1 = pd.Categorical(df['score'])
print(ct1.dtype)
print('''使用 Categorical 类将 df['score'] 转换为 Categorical 类型''')
print(ct1.codes)
# 输出 Categorical 类型中['a', 'b', 'b', 'b']的分类统计信息
print('''输出 Categorical 类型中['a', 'b', 'b', 'b']的分类统计信息''')
```

输出如下:

```
name      category
score     category
gender    category
dtype: object
使用 astype 函数将 df 转换为 Categorical 类型
category
使用 Categorical 类将 df['score'] 转换为 Categorical 类型
[0 1 1 1]
输出 Categorical 类型中['a', 'b', 'b', 'b']的分类统计信息
```

4.2.11 Pandas IO 操作

Pandas 提供了读写 Excel 文件和 csv 格式文件的函数,见表 4-7。

表 4-7 Pandas IO 操作

函 数 名 称	描 述
ExcelFile	读取 Excel 文件
to_excel	写入 Excel 文件
read_csv	读取 csv 格式文件
to_csv	写入 csv 格式文件

【实例 4-16】 使用 Pandas 读取和写入 Excel 文件和 csv 格式文件。

```
import pandas as pd
```

```
    # 使用 ExcelFile 函数读取 excel_read.xlsx 文件
    data_excel_read = pd.ExcelFile('excel_read.xlsx')
    # 读取 excel_read.xlsx 的数据
    print(data_excel_read.parse(0))
    print('使用 ExcelFile 函数读取 excel_read.xlsx 文件')
    # 使用 DataFrame 类转换 data_excel_read 后，使用 to_excel 函数写入 excel_write.xlsx 文件
    pd.DataFrame(data_excel_read.parse(0)).to_excel('excel_write.xlsx',sheet_name='商品信息',index=False,header=True)
    print('使用 DataFrame 类转换 data_excel_read 后，使用 to_excel 函数写入 excel_write.xlsx 文件')

    # 使用 read_csv 函数读取 csv_read.csv 文件
    data_csv_read = pd.read_csv('csv_read.csv')
    print(data_csv_read)
    print('使用 read_csv 函数读取 csv_read.csv 文件')
    # 使用 DataFrame 类转换 data_csv_read 后，使用 to_csv 函数写入 csv_write.csv 文件
    pd.DataFrame(data_csv_read).to_csv('csv_write.csv',index=False,header=True)
    print('使用 DataFrame 类转换 data_csv_read 后，使用 to_csv 函数写入 csv_write.csv 文件')
```

输出如下：

```
       类别       细分        订单 Id           订单日期
    0  家具      消费者     CN-2014-1190387  2015-02-28
    1  办公用品   小型企业    CN-2012-2932548  2013-05-17
    2  办公用品   小型企业    CN-2015-3665640  2016-10-23
    3  办公用品   消费者     CN-2015-3232413  2016-12-04
    4  家具      消费者     CN-2015-3381572  2016-05-11
    5  技术      公司      CN-2014-2662386  2015-03-04
    6  办公用品   公司      US-2013-1498643  2014-11-26
    7  办公用品   公司      CN-2014-3807036  2015-01-21
    8  办公用品   公司      US-2014-3678373  2015-08-28
    使用 ExcelFile 函数读取 excel_read.xlsx 文件
    使用 DataFrame 类转换 data_excel_read 后，使用 to_excel 函数写入 excel_write.xlsx 文件
       姓名        班级           成绩
    0  Simon     Class one     100
    1  Tom       Class two     87
    2  Joy       Class three   90
    3  Alex      Class four    96
    使用 read_csv 函数读取 csv_read.csv 文件
    使用 DataFrame 类转换 data_csv_read 后，使用 to_csv 函数写入 csv_write.csv 文件
```

【任务实施】

任务 1　使用 Pandas 实现水果销售数据分析

使用 Pandas 实现某超市水果销售数据分析，包括数据获取、数据清洗、数据建模、数据可视化和数据结果分析。统计水果每月消费次数、月均消费金额、客户人均消费金额，使用柱状图呈现实收金额和时间的关系，使用条形图呈现和分析水果每月销售金额，统计并使用饼图展示销售数量前 10 的水果。

查看 Excel 源数据，了解数据概况。源数据包含 7 个字段：出售时间、客户 ID、产品 ID、水果名、购买斤数、消费总额和实收金额，如图 4-4 所示。

出售时间	客户ID	产品ID	水果名	购买斤数	消费总额	实收金额
2020-01-01 星期五	0016165299	6701	百香果	6	84	78
2020-01-02 星期六	001616528	6701	百香果	1	14	13.5
2020-01-06 星期三	0012602828	6701	百香果	2	28	28
2020-01-11 星期一	0010070343428	6701	百香果	2	28	25.55
2020-01-15 星期五	00101554328	6701	百香果	8	112	109
2020-01-20 星期三	0013389528	6701	百香果	1	14	14
2020-01-31 星期日	00101464928	6701	百香果	3	42	40.51
2020-02-17 星期一	0011177328	6702	阳光玫瑰葡萄	5	149	131.12
2020-02-22 星期一	0010065687828	6702	阳光玫瑰葡萄	1	29.8	26.22
2020-02-24 星期三	0013389528	6702	阳光玫瑰葡萄	4	119.2	104.89
2020-03-05 星期六	0010026389628	6702	阳光玫瑰葡萄	2	59.6	59.6
2020-03-05 星期六	00102285028	6702	阳光玫瑰葡萄	3	84	84
2020-03-05 星期六	0010077400828	6702	阳光玫瑰葡萄	1	28	24.64
2020-03-07 星期一	0010077400828	6702	阳光玫瑰葡萄	5	140	112
2020-03-09 星期三	0010079843728	6702	阳光玫瑰葡萄	6	168	140
2020-03-15 星期二	0010031328528	6702	阳光玫瑰葡萄	2	56	49.28
2020-03-15 星期二	00100703428	6702	阳光玫瑰葡萄	2	56	49.28
2020-03-15 星期二	0010712328	6702	阳光玫瑰葡萄	5	140	112

图 4-4 查看 Excel 源数据（部分）

1）使用 Pandas 获取 Excel 文件数据，并将数据统一转换为 object 类型。由于在源数据中可能存在数据类型不一致的情况，存在难以被自动区分的类型，而 Python 中一切皆对象，因此将它们都处理为最宽泛的"对象"，也就是 object 类型，这样可以保证顺利获取数据。

```
# 导入 Pandas
import pandas as pd
# 使用 object 类型读取数据
sale_data = pd.read_excel("超市水果销售数据.xlsx", dtype="object")
# 使用 info 函数查看 sale_data 数据类型
sale_data.info()
```

输出如图 4-5 所示。

2）将获取的数据转换为 Pandas 的 DataFrame 类型，并查看相关属性：shape、index 和 columns。

```
# 转换为 DataFrame 格式
df_result = pd.DataFrame(sale_data)
# 查看源数据前面 10 行数据
df_result.head(10)
```

输出如图 4-6 所示。

```
<class 'pandas.core.frame.DataFrame'>
RangeIndex: 6287 entries, 0 to 6286
Data columns (total 7 columns):
 #   Column    Non-Null Count  Dtype
---  ------    --------------  -----
 0   出售时间      6283 non-null   object
 1   客户ID      6283 non-null   object
 2   产品ID      6286 non-null   object
 3   水果名       6286 non-null   object
 4   购买斤数      6286 non-null   object
 5   消费总额      6286 non-null   object
 6   实收金额      6286 non-null   object
dtypes: object(7)
memory usage: 343.9+ KB
```

图 4-5 sale_data 数据基本信息

	出售时间	客户ID	产品ID	水果名	购买斤数	消费总额	实收金额
0	2020-01-01 星期五	0016165299	6701	百香果	6	84	78
1	2020-01-02 星期六	001616528	6701	百香果	1	14	13.5
2	2020-01-06 星期三	0012602828	6701	百香果	2	28	28
3	2020-01-11 星期一	0010070343428	6701	百香果	2	28	25.55
4	2020-01-15 星期五	00101554328	6701	百香果	8	112	109
5	2020-01-20 星期三	0013389528	6701	百香果	1	14	14
6	2020-01-31 星期日	00101464928	6701	百香果	3	42	40.51
7	2020-02-17 星期一	0011177328	6702	阳光玫瑰葡萄	5	149	131.12
8	2020-02-22 星期一	0010065687828	6702	阳光玫瑰葡萄	1	29.8	26.22
9	2020-02-24 星期三	0013389528	6702	阳光玫瑰葡萄	4	119.2	104.89

图 4-6 源数据前 10 行数据

```
# 查看数据的形状,即几行几列
df_result.shape
```

输出如下:

```
(6287, 7)
```

```
# 查看索引
df_result.index
```

输出如下:

```
RangeIndex(start=0, stop=6287, step=1)
```

```
# 查看每一列的列表头内容
df_result.columns
```

输出如下:

```
Index(['出售时间', '客户ID', '产品ID', '水果名', '购买斤数', '消费总额', '实收金额'], dtype='object')
```

```
# 查看每一列数据统计数目
df_result.count()
```

输出如下:

```
出售时间      6283
客户ID       6283
产品ID       6286
水果名        6286
购买斤数       6286
消费总额       6286
实收金额       6286
dtype: int64
```

3)通过观察 DataFrame 数据的基本信息和属性值,发现源数据中总共包含 6286 行数据,7 个字段:出售时间、客户 ID、产品 ID、水果名、购买斤数、消费总额和实收金额。其中出售时间、客户 ID 两个字段分别只有 6283 行数据,其余字段各有 6286 行数据,因此可以判断其中出售时间、客户 ID 两个字段存在缺失值,接下来将字段"出售时间"修改为"销售日期"。

```
# 使用 rename 函数,把"出售时间"改为"销售日期"
df_result.rename(columns={"出售时间": "销售日期"}, inplace=True)
df_result.columns
```

输出如下:

```
Index(['销售日期', '客户ID', '产品ID', '水果名', '购买斤数', '消费总额', '实收金额'], dtype='object')
```

```
# 使用 dropna 函数删除缺失值
df_result = df_result.dropna()
# 删除缺失值之后
df_result.shape
```

输出如下:

```
(6280, 7)
```

```
# 查看处理缺失值后的结果
df_result.isnull()
```

输出如图 4-7 所示。

	销售日期	客户ID	产品ID	水果名	购买斤数	消费总额	实收金额
0	False	False	False	False	False	False	False
1	False	False	False	False	False	False	False
2	False	False	False	False	False	False	False
3	False	False	False	False	False	False	False
4	False	False	False	False	False	False	False
...	...	...	...	...	...	...	...
6281	False	False	False	False	False	False	False
6282	False	False	False	False	False	False	False
6284	False	False	False	False	False	False	False
6285	False	False	False	False	False	False	False
6286	False	False	False	False	False	False	False

6280 rows × 7 columns

图 4-7　处理缺失值后的结果

4）由于之前为了成功导入数据，将数据类型全部转换为 object 类型了。这里为了更好地符合数据分析的实际需求，需要对部分数据类型进行转换。例如，销售日期转换为时间类型，购买斤数、消费总额和实收金额转换为浮点数类型。

```
# 将字符串转为浮点型数据
df_result["购买斤数"] = df_result["购买斤数"].astype("f8")
df_result["消费总额"] = df_result["消费总额"].astype("f8")
df_result["实收金额"] = df_result["实收金额"].astype("f8")
df_result.dtypes
```

输出如下：

```
销售日期      object
客户ID      object
产品ID      object
水果名       object
购买斤数      float64
消费总额      float64
实收金额      float64
dtype: object
```

```
# 查看销售日期字段
df_result["销售日期"]
```

输出如下：

```
销售日期
0    2020-01-01 星期五
1    2020-01-02 星期六
2    2020-01-06 星期三
3    2020-01-11 星期一
```

```
4         2020-01-15 星期五
                 ⋮
6281      2020-04-27 星期三
6282      2020-04-27 星期三
6284      2020-04-27 星期三
6285      2020-04-27 星期三
6286      2020-04-28 星期四
Name: 销售日期, Length: 6280, dtype: object
# 定义函数将星期去除
def splitsaleweek(timeColser):
    datelist = []
    for t in timeColser:
        datelist.append(t.split(" ")[0])   # [0]表示选取的分片,这里表示切割完后选取第一个分片
    timeser = pd.Series(datelist)   # 将列表转行为一维数据 Series 类型
    return timeser

# 获取"销售日期"这一列数据
t = df_result.loc[:, "销售日期"]
# 调用函数去除星期,获取日期
timeser = splitsaleweek(t)
# 修改"销售日期"这一列日期
df_result.loc[:, "销售日期"] = timeser
df_result.head(10)
```

输出如图 4-8 所示。

	销售日期	客户ID	产品ID	水果名	购买斤数	消费总额	实收金额
0	2020-01-01	0016165299	6701	百香果	6.0	84.0	78.00
1	2020-01-02	001616528	6701	百香果	1.0	14.0	13.50
2	2020-01-06	0012602828	6701	百香果	2.0	28.0	28.00
3	2020-01-11	0010070343428	6701	百香果	2.0	28.0	25.55
4	2020-01-15	00101554328	6701	百香果	8.0	112.0	109.00
5	2020-01-20	0013389528	6701	百香果	1.0	14.0	14.00
6	2020-01-31	00101464928	6701	百香果	3.0	42.0	40.51
7	2020-02-17	0011177328	6702	阳光玫瑰葡萄	5.0	149.0	131.12
8	2020-02-22	0010065687828	6702	阳光玫瑰葡萄	1.0	29.8	26.22
9	2020-02-24	0013389528	6702	阳光玫瑰葡萄	4.0	119.2	104.89

图 4-8 去除星期数据的 df_result

```
# 字符串转日期
# errors='coerce'如果原始数据不符合日期的格式,转换后的值为 NaT
df_result.loc[:, "销售日期"] = pd.to_datetime(df_result.loc[:, "销售日期"], errors='coerce')
df_result.dtypes
```

输出如下:

```
销售日期          datetime64[ns]
客户 ID         object
产品 ID         object
水果名           object
购买斤数          float64
```

```
消费总额           float64
实收金额           float64
dtype: object
```

```python
# 转换日期过程中不符合日期格式的数值会被转换为空值 None
# 这里删除为空的行
df_result = df_result.dropna()
df_result.shape
```

输出如下：

```
(6275, 7)
```

5）根据时间进行升序排序，同时将索引也进行重置。

```python
# 按销售日期进行升序排序
df_result = df_result.sort_values(by='销售日期', ascending=True)
df_result.head()
```

输出如图 4-9 所示。

	销售日期	客户ID	产品ID	水果名	购买斤数	消费总额	实收金额
0	2020-01-01	0016165299	6701	百香果	6.0	84.0	78.0
868	2020-01-01	0013331728	1405	泰国山竹	2.0	69.0	62.0
867	2020-01-01	0011743428	1405	泰国山竹	1.0	34.5	31.0
1856	2020-01-01	0010060654328	1458	京红桃	1.0	10.3	9.2
2130	2020-01-01	00103283128	1464	鸭梨	1.0	2.5	2.2

图 4-9　按销售日期排序

```python
# 重置索引（index）
df_result = df_result.reset_index(drop=True)
df_result.head()
```

输出如图 4-10 所示。

	销售日期	客户ID	产品ID	水果名	购买斤数	消费总额	实收金额
0	2020-01-01	0016165299	6701	百香果	6.0	84.0	78.0
1	2020-01-01	0013331728	1405	泰国山竹	2.0	69.0	62.0
2	2020-01-01	0011743428	1405	泰国山竹	1.0	34.5	31.0
3	2020-01-01	0010060654328	1458	京红桃	1.0	10.3	9.2
4	2020-01-01	00103283128	1464	鸭梨	1.0	2.5	2.2

图 4-10　重置索引

6）查看数据的描述统计信息后发现，当前数据字段：购买斤数、消费总额和实收金额的最小值存在负数。这显然是不合理的，因此需要对这类数据进行异常值处理。同时进行重复值处理。

```python
# 查看描述统计信息
df_result.describe()
```

输出如图 4-11 所示。

```python
# 将"购买斤数"这一列中小于 0 的数排除掉
pop = df_result.loc[:, "购买斤数"] > 0
df_result = df_result.loc[pop, :]
```

```
# 删除重复数据
df_kpi1 = df_result.drop_duplicates(subset=['销售日期', '客户ID'])
# 排除异常值后再次查看描述统计信息
df_result.describe()
```

输出如图 4-12 所示。

	购买斤数	消费总额	实收金额
count	6275.000000	6275.000000	6275.000000
mean	2.392032	50.416239	46.355624
std	2.323516	74.750866	71.724141
min	-10.000000	-374.000000	-374.000000
25%	1.000000	14.000000	12.600000
50%	2.000000	29.000000	27.440000
75%	2.000000	59.600000	54.000000
max	33.000000	1372.000000	1372.000000

图 4-11　存在异常值的描述统计信息

	购买斤数	消费总额	实收金额
count	6234.000000	6234.000000	6234.000000
mean	2.413539	50.898653	46.803065
std	2.311527	74.591501	71.577203
min	1.000000	1.200000	0.030000
25%	1.000000	14.000000	12.600000
50%	2.000000	29.800000	28.000000
75%	2.000000	59.600000	54.300000
max	33.000000	1372.000000	1372.000000

图 4-12　排除异常值后的描述统计信息

7）统计水果每月消费次数，即每月消费次数=总消费次数/月份数。

```
# 总消费次数
total = df_kpi1.shape[0]
print('总消费次数: ', total)
```

输出如下：

总消费次数： 5360

```
# 按销售时间升序排序
df_kpi1 = df_kpi1.sort_values(by='销售日期', ascending=True)
# 重命名行名(index)
df_kpi1 = df_kpi1.reset_index(drop=True)
df_kpi1
```

输出如图 4-13 所示。

	销售日期	客户ID	产品ID	水果名	购买斤数	消费总额	实收金额
0	2020-01-01	0016165299	6701	百香果	6.0	84.0	78.0
1	2020-01-01	001616528	1456	甘蔗	1.0	7.0	7.0
2	2020-01-01	00101470528	6709	美早大樱桃	4.0	179.2	159.2
3	2020-01-01	0010074557328	67010	高乐密芒果	1.0	5.6	5.6
4	2020-01-01	0013401428	1405	泰国山竹	1.0	34.5	31.0
...	...	...	...	...	...	...	...
5355	2020-07-19	0010019136328	5099	丑橘	2.0	2.4	2.0
5356	2020-07-19	0100013306428	67011	桂花香荔枝	1.0	31.0	28.0
5357	2020-07-19	0010065621228	1435	酥梨	2.0	71.6	64.0
5358	2020-07-19	0010030713328	67011	桂花香荔枝	4.0	124.0	118.0
5359	2020-07-19	0010542828	5425	墨童西瓜	6.0	69.0	57.5

5360 rows × 7 columns

图 4-13　按销售日期升序排并重置索引

```
# 获取时间范围
# 最小时间值
startDate = df_kpi1.loc[0, '销售日期']
# 最大时间值
endDate = df_kpi1.loc[total - 1, '销售日期']
# 计算天数
days_1 = (endDate - startDate).days
# 月份数：运算符"//"表示取整除，返回商的整数部分
months_1 = days_1 // 30
print('月份数: ', months_1)
```

输出如下：

月份数: 6

```
# 计算月均消费次数
kpi1_1 = total // months_1
print('业务指标1：月均消费次数=', kpi1_1)
```

输出如下：

业务指标1：月均消费次数= 893

8）统计每月平均消费金额，即每月平均消费金额=总消费金额/月份数。

```
# 总消费金额
Totalconsumption = df_result.loc[:, '实收金额'].sum()
Totalconsumption
```

输出如下：

291770.31

```
# 月均消费金额
monthconsumption = Totalconsumption / months_1
print('业务指标2：月均消费金额=', monthconsumption)
```

输出如下：

业务指标2：月均消费金额= 48628.385

9）统计客户人均消费金额，即消费总金额/消费总次数。

```
# 客户人均消费金额 = 消费总金额 / 消费总次数
pcc = Totalconsumption / total
print('业务指标3：客户人均消费金额=', pcc)
```

输出如下：

业务指标3：客户人均消费金额= 54.43475932835821

10）以可视化方式呈现和分析水果每天销售趋势，为了方便可视化显示中文字体，这里需要导入相应的字体包。

```
import matplotlib as mpl
# 汉字字体，优先使用楷体，如果找不到楷体，则使用黑体
mpl.rcParams['font.sans-serif'] = ['KaiTi', 'SimHei', 'FangSong']
# 字体大小
mpl.rcParams['font.size'] = 12
# 正常显示负号
```

```
mpl.rcParams['axes.unicode_minus'] = False
# 在操作之前先复制一份数据，防止影响清洗后的数据
df_result1 = df_result
# 重命名行（index）为销售日期所在列的值
df_result1.index = df_result1['销售日期']
df_result1.head(10)
```

输出如图 4-14 所示。

销售日期	销售日期	客户ID	产品ID	水果名	购买斤数	消费总额	实收金额
2020-01-01	2020-01-01	0016165299	6701	百香果	6.0	84.0	78.00
2020-01-01	2020-01-01	0013331728	1405	泰国山竹	2.0	69.0	62.00
2020-01-01	2020-01-01	0011743428	1405	泰国山竹	1.0	34.5	31.00
2020-01-01	2020-01-01	0010060654328	1458	京红桃	1.0	10.3	9.20
2020-01-01	2020-01-01	00103283128	1464	鸭梨	1.0	2.5	2.20
2020-01-01	2020-01-01	0012697828	1464	鸭梨	4.0	10.0	9.40
2020-01-01	2020-01-01	0011811728	1492	雪花梨	1.0	20.0	18.00
2020-01-01	2020-01-01	0010059484328	67012	新西兰金果	2.0	49.0	43.12
2020-01-01	2020-01-01	0013448228	1507	国产香蕉	1.0	9.5	8.50
2020-01-01	2020-01-01	0010616728	5099	丑橘	2.0	3.4	3.00

图 4-14　df_result1 副本数据信息

```
import matplotlib.pyplot as plt
# 使用柱状图呈现实收金额和日期的关系
plt.plot(df_result1['实收金额'],color='y')
plt.title('按天消费金额图')
plt.xlabel('时间')
plt.ylabel('实收金额')
```

输出如图 4-15 所示。

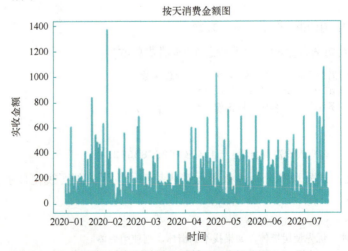

图 4-15　每天消费金额可视化柱状图

分析结果：从每天消费金额可视化柱状图中可以发现，每天消费金额主要分布在 300～500 元这个水平。

11）以可视化方式呈现和分析水果每月销售金额，统计销售数量排名前 10 的水果，使用饼图展示销售数量前 10 的水果。

```
# 将销售日期聚合按月分组
month_group = df_result1.groupby(df_result1.index.month)
# 应用函数，计算每个月的消费总额
df_month = month_group.sum()
df_month
```

输出如图 4-16 所示。

```
# 描绘按月消费金额图
plt.bar(df_month.index,df_month['实收金额'],color='r')
plt.title('按月消费金额图')
plt.xlabel('月份')
plt.ylabel('实收金额')
```

输出如图 4-17 所示。

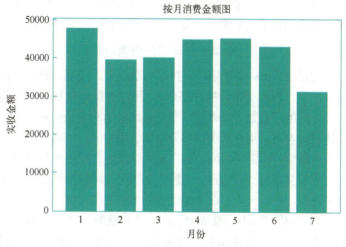

图 4-16　每月销售总额　　　　图 4-17　每月消费金额可视化柱状图

分析结果：从每月消费金额可视化柱状图中可以发现，由于 7 月数据不完整所以不具备参考性，1~6 月数据大小不一，可能是由于 2、3 月处于农历新年期间，购买水果的需求不高。

12）统计销售数量排名前 10 的水果。

```
# 聚合统计各种水果的销售数量
fruits = df_result1[['水果名','购买斤数']]
fruits_1 = fruits.groupby('水果名')[['购买斤数']]
re_fruits = fruits_1.sum()
# 对水果购买数量按降序排序
re_fruits = re_fruits.sort_values(by='购买斤数',ascending=False)
# 截取销售数量最多的 10 种水果
top_fruits = re_fruits.iloc[:10,:]
top_fruits
```

输出如图 4-18 所示。

```
# 用饼图展示销售数量前 10 的水果
top10_fruits = top_fruits['购买斤数']
top10_fruits.plot(kind='pie',autopct='%1.1f%%',title='水果销售前 10 情况')
```

输出如图 4-19 所示。

图4-18 销售前10的水果　　　　图4-19 销售前10的水果饼图

分析结果：从销售前 10 的水果饼图中可以发现，桂花香荔枝、国产香蕉和甘蔗比较畅销。这可能跟当地人的喜好有关。

任务 2　使用 Pandas 实现用户消费行为数据分析

使用 Pandas 实现用户消费行为数据分，包括数据清洗操作（缺失值、去重、数据类型转换和整合）、每月消费总金额、每月购买产品总数、每月用户总数、每月用户数量对比、用户消费金额与购买产品数量描述和散点图、用户消费次数分布图、用户累计消费金额占比、用户第一次和最后一次消费时间、用户分类以及老客户的回购率统计。

查看文本格式源数据，了解数据概况。源数据包含 4 列数据，但没有字段名，根据分析将在下面赋予适合的字段名称，如图 4-20 所示。

1）使用 Pandas 的 read_table 函数获取文本格式数据，通过之前的数据观察，发现源数据的分隔符不是逗号，因此使用的列分隔符为 sep = '\s+'（s+可以将 tab 和多个空格都当成一样的分隔）。同时，给该数据源增加字段名称['顾客 ID','购买日期','购买产品数量','消费金额']。

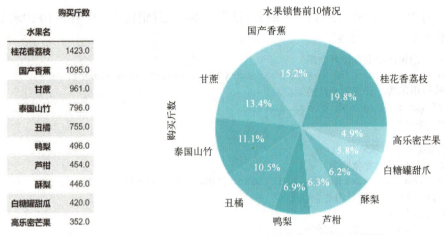

图4-20　文本格式源数据

```
import pandas as pd
import numpy as np
import matplotlib.pyplot as plt
import matplotlib as mpl
# 汉字字体,优先使用楷体,如果找不到楷体,则使用黑体['KaiTi', 'SimHei', 'FangSong']
mpl.rcParams['font.sans-serif'] = ['KaiTi', 'SimHei', 'FangSong']
# 字符集选择
# mpl.rcParams['font.family'] = 'AR PL UKai CN '
# 字体大小
mpl.rcParams['font.size'] = 12
# 正常显示负号
cols = ['顾客ID','购买日期','购买产品数量','消费金额']
# 顾客ID 购买日期 购买产品数量 消费金额
df_data = pd.read_table("客户数据消费.txt",names = cols,sep = '\s+')
```

2）使用 info 查看 df_data 数据是否存在空值及相应的数据类型。使用 head 函数查看 df_data 数据内容前 5 行。这里的"购买日期"的数据类型是 int64，因此后面需要将其转换为日期格式。

```
# 使用 info 查看是否存在空值
print(df_data.info())
# 查看前 5 行数据
print(df_data.head())
```

输出如图 4-21 和图 4-22 所示。

```
<class 'pandas.core.frame.DataFrame'>
RangeIndex: 66231 entries, 0 to 66230
Data columns (total 4 columns):
 #   Column      Non-Null Count  Dtype
---  ------      --------------  -----
 0   顾客ID        66231 non-null  int64
 1   购买日期        66231 non-null  int64
 2   购买产品数量      66231 non-null  int64
 3   消费金额        66231 non-null  float64
dtypes: float64(1), int64(3)
memory usage: 2.0 MB
None
```

	顾客ID	购买日期	购买产品数量	消费金额
0	1	20210101	1	11.77
1	2	20210112	1	12.00
2	3	20210102	2	20.76
3	3	20210330	2	20.76
4	3	20200528	1	16.99

图 4-21　查看 df_data 基本信息　　　　　图 4-22　查看 df_data 数据内容

3）将获取的数据 df_data['购买日期']进行部分时间格式类型转换，并根据 df_data.购买日期新增一列 df_data['月份']数据，类型为 datetime64[M]。查看 df_data 的描述统计信息，可以看到平均的购买产品数量是 2.4，中位数是 2，最小值是 1，最大值是 99。平均消费金额是 35.87，中位数是 36，最小值是 0，最大值是 1286。最大值和最小值的差别比较大，存在极值干扰。

```
# 转换购买日期字段的数据类型为时间类型，format ='%Y%m%d'表示年月日。
df_data['购买日期'] = pd.to_datetime(df_data.购买日期,format ='%Y%m%d')
# 在 df_data 里增加一列：月份，并将其类型设置为日期类型
df_data['月份'] = df_data.购买日期.values.astype('datetime64[M]')
# 去掉时分秒
df_data['月份'] = df_data['月份'].dt.date
# 查看当前 df 的数据内容
df_data
```

输出如图 4-23 所示。

```
# 查看当前 df 的数据描述信息
df_data.describe()
```

输出如图 4-24 所示。

	顾客ID	购买日期	购买产品数量	消费金额	月份
0	1	2021-01-01	1	11.77	2021-01-01
1	2	2021-01-12	1	12.00	2021-01-01
2	3	2021-01-02	2	20.76	2021-01-01
3	3	2021-03-30	2	20.76	2021-03-01
4	3	2020-05-28	1	16.99	2020-05-01
...	...	...	...	...	...
66226	23414	2021-03-25	1	11.77	2021-03-01
66227	23414	2021-04-01	1	14.96	2021-04-01
66228	23414	2021-05-12	2	28.74	2021-05-01
66229	23414	2021-11-07	2	27.48	2021-11-01
66230	23414	2021-12-27	1	5.49	2021-12-01

66231 rows × 5 columns

	顾客ID	购买产品数量	消费金额
count	66231.000000	66231.000000	66231.000000
mean	11491.160438	2.409461	35.870957
std	6689.833168	2.322714	36.012428
min	1.000000	1.000000	0.000000
25%	5765.000000	1.000000	14.490000
50%	11470.000000	2.000000	25.980000
75%	17052.000000	3.000000	43.690000
max	23414.000000	99.000000	1286.010000

图 4-23　df 的数据内容　　　　　图 4-24　df_data 的数据描述信息

4）使用 groupby 函数按月分析销售数据：每月消费总金额、每月购买产品总数和每月用户总数。分别使用柱状图、折线图、饼图表示出来。最后，对每月用户总数进行去重。

```python
# 按月分析数据趋势
# 按月分组
gp_month = df_data.groupby('月份')
# 使用 sum 函数计算每月消费总金额
month_spend = gp_month.消费金额.sum()
# 使用 sum 函数计算每月购买产品总数
month_total = gp_month.购买产品数量.sum()
# 使用 count 函数计算每月用户总数
month_persons = gp_month.顾客ID.count()
# 使用 plot 函数绘制 month_spend 的柱状图, month_total 的折线图, month_persons 的饼图数据
# 每月消费总金额
month_spend.plot(kind="bar",color="red")
# 每月购买产品总数
month_total.plot(color="blue")
# 每月用户百分比
month_persons.plot(kind="pie",autopct='%1.0f%%')
plt.show()
# 每月消费人数,使用 groupby 函数根据月份字段进行分组,nunique 表示去重
df_data.groupby('月份').顾客ID.nunique().plot()
plt.show()
```

输出如图 4-25～图 4-28 所示。

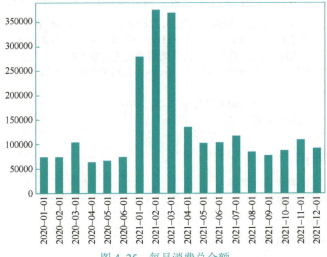

图 4-25　每月消费总金额

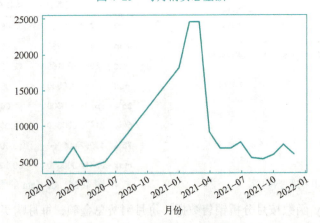

图 4-26　每月购买产品总数

项目 4　Pandas 数据分析和处理

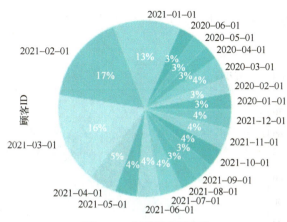

图 4-27　每月用户百分比

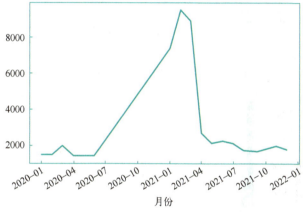

图 4-28　去重后的每月消费人数

5）查看数据的统计描述信息，可以发现根据顾客 ID 分组之后，每个用户平均购买产品的数量是 7.1，中位数是 17，最小值是 1，最大值是 1033。每个用户的平均消费金额是 105.99，中位数是 243.16，最小值是 0，最大值是 13990.93。

```
# 单个客户消费数据分析
# 使用 groupby 根据顾客 ID 进行分组
gp_cusID = df_data.groupby('顾客 ID')
# 查看将 gp_cusID 的数据求和后的描述信息
print(gp_cusID.sum().describe())
```

输出如图 4-29 所示。

图 4-29　将 gp_cusID 的数据求和后的描述信息

6）根据顾客 ID 进行分组，可视化统计消费金额小于 1000 的散点图，以及购买产品小于 30 的直方图。

```
# 使用 groupby 根据顾客 ID 进行分组
gp_cusID = df_data.groupby('顾客 ID')
# 使用 sum 函数求和之后，使用 query 函数查询消费金额<1000 的数据，并使用 scatter 函数将所得数据的消费金额字段作为 x 轴，购买产品数量作为 y 轴。
gp_cusID.sum().query("消费金额<1000").plot.scatter(x = '消费金额',y = '购买产品数量')
# 使用 sum 函数根据顾客 ID 求和之后，根据 query 函数查询购买产品数量<30 的数据，并绘制直方图，bins = 40 表示分成 40 组数据
```

```
gp_cusID.sum().query("购买产品数量<30").购买产品数量.plot.hist(bins = 40)
# 显示直方图
plt.show()
```

输出如图 4-30 和图 4-31 所示。

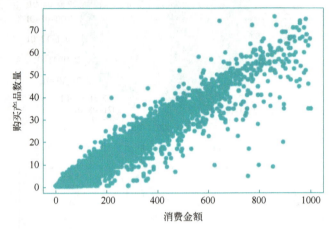

图 4-30 消费金额小于 1000 的散点图

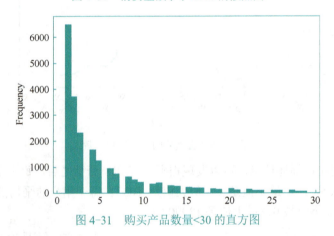

图 4-31 购买产品数量<30 的直方图

7）计算用户的累计消费金额占比。

```
# 用户的累计消费金额占比
# 根据顾客 ID 求和之后，按照消费金额字段进行排序，cumsum 方法是滚动求和
cum1=gp_cusID.sum().sort_values("消费金额").apply(lambda x:x.cumsum()/x.sum())
# 使用 reset_index 函数对排序之后的数据索引进行重置
cum1.reset_index().消费金额.plot()
plt.show()
```

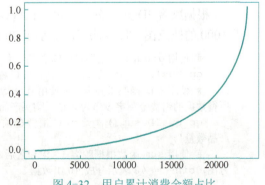

输出如图 4-32 所示。从图 4-32 中可以看出，超过一大半消费用户的消费金额还不到总消费金额的 20%，因此，剩余的小部分大客户是需要重点关照的对象。

8）用户消费行为分析，统计用户第一次和最后一次消费的时间所集中的月份和次数对比。

```
# 用户消费行为分析
```

图 4-32 用户累计消费金额占比

```python
# 根据顾客ID提取月份的min最小值：表示用户第一次消费的时间所集中的月份和次数
print(gp_cusID.月份.min().value_counts())
# 根据顾客ID提取月份的max最大值：表示用户最后一次消费的时间所集中的月份和次数
print(gp_cusID.月份.max().value_counts())
gp_cusID.月份.min().value_counts().plot()
gp_cusID.月份.max().value_counts().plot()
```

输出如图 4-33 和图 4-34 所示。从图中可以发现用户第一次和最后一次消费时间多是在 2021 年的 1~3 月份。很多用户只有一次购买记录。

```
2021-02-01    6474
2021-01-01    5711
2021-03-01    5122
2020-01-01    1460
2020-03-01    1154
2020-02-01    1016
2020-04-01     577
2020-06-01     453
2020-05-01     447
2021-08-01       1
Name: 月份, dtype: int64
2021-02-01    5302
2021-03-01    4645
2021-01-01    4219
2021-12-01    1769
2021-11-01    1331
2021-10-01     911
2021-04-01     816
2021-07-01     776
2021-06-01     705
2021-09-01     684
2021-08-01     636
2021-05-01     616
2020-03-01       2
2020-06-01       2
2020-04-01       1
Name: 月份, dtype: int64
```

图 4-33 用户第一次和最后一次消费的时间所集中的月份和次数

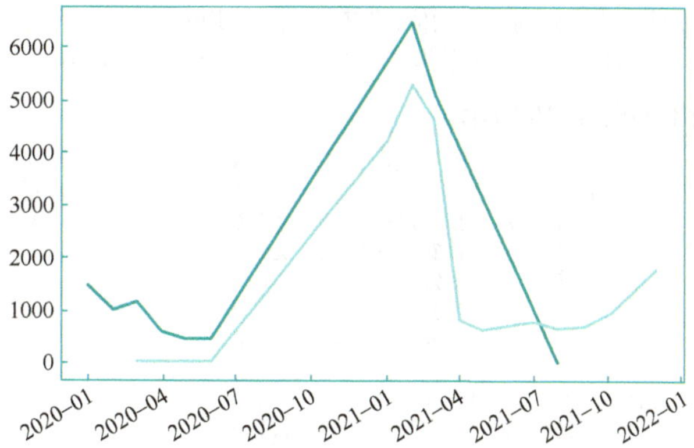

图 4-34 用户第一次和最后一次消费的时间所集中的月份和次数折线图

9）统计用户第一次和最后一次购买时间，发现有 11470 个用户只购买了一次，10945 个用户有重复购买记录。

```
# 使用聚合函数agg，根据gp_cusID的购买日期字段求出用户第一次和最后一次购买时间
new = gp_cusID.购买日期.agg(['min','max'])
# 如果 new['min']==new['max']则表示：该用户第一次和最后一次购买是同一时间，就可能是
流失的客户数
print((new['min']==new['max']).value_counts())
exit()
```

输出如下：

```
True     11470
False    10945
dtype: int64
```

10）下面将开始对数据做透视表处理，并修改其中的字段名，转换部分数据类型。

```
# 使用 pivot_table 函数实现 df 的数据透视表功能，index 表示索引，values 表示字段，
aggfunc表示聚合函数(max:求最大值,sum:求和)
```

```
df_rfm = df_data.pivot_table(index = '顾客ID',
                    values = ['购买产品','消费金额','购买日期'],
                    aggfunc = {'购买日期':'max','消费金额':'sum','购买产品':'sum'})
# 查看 rfm 的数据透视表
# df_rfm
# 将购买日期与最后一次购买日期相减，再除以按天（D）计算的时间格式，得到用户最近一次交易
时间间隔 rfm["R"]
df_rfm["R"] = -(df_rfm.购买日期-df_rfm.购买日期.max())/np.timedelta64(1,'D')
# 使用 rename 函数将消费金额重命名为 M，购买产品数量重命名为 F
df_rfm.rename(columns = {'消费金额':'M','购买产品数量':'F'},inplace=True)
print(df_rfm)
exit()
```

输出如图 4-35 所示。

11）根据上面的数据透视表，可以通过分别计算 R、F、M 各自与平均值相减后的正负值，表示该行数据在整个数据中的位置，目的是将该行数据所代表的用户在将来进行分类标识。

```
# 使用 apply(lambda x: x - x.mean())求出每个用户 ID 的 R、F、M 各自与平均值相减后的
正负值，正表示高于平均值，负表示低于平均值
print(df_rfm[["R", 'F', 'M']].apply(lambda x: x - x.mean()))
```

输出如图 4-36 所示。

```
           M   F   购买日期        R
顾客ID
1       11.77   1  2021-01-01  364.0
2       12.00   1  2021-01-12  353.0
3       58.51   5  2021-03-30  276.0
4      100.50   7  2021-12-12   19.0
5      167.70  13  2021-12-12   19.0
 ⋮        ⋮    ⋮      ⋮          ⋮
23356  216.31  12  2021-11-22   39.0
23357  154.90  13  2021-07-07  177.0
23358   23.54   2  2021-03-24  282.0
23413   87.93   7  2021-12-04   27.0
23414   88.44   7  2021-12-27    4.0

[22415 rows x 4 columns]
```

```
             R         F          M
顾客ID
1         120.4 -6.119384 -94.220156
2         109.4 -6.119384 -93.990156
3          32.4 -2.119384 -47.480156
4        -224.6 -0.119384  -5.490156
5        -224.6  5.880616  61.709844
 ⋮          ⋮       ⋮          ⋮
23356    -204.6  4.880616 110.319844
23357     -66.6  5.880616  48.909844
23358      38.4 -5.119384 -82.450156
23413    -216.6 -0.119384 -18.060156
23414    -239.6 -0.119384 -17.550156

[22415 rows x 3 columns]
```

图 4-35　查看 df_rfm 数据透视表　　　图 4-36　R、F、M 各自与平均值相减后的正负值

12）自定义函数 aggfc，用于根据 R、F、M 的正负值对其用户进行编码处理：正数用 1 表示，负数用 0 表示。这里将用户分为 8 个类型。通过增加新列 rfm['label']来对 8 类用户进行求和与个数统计操作，并使用散点图显示 2 类客户和非 2 类客户的分布情况。

```
# 自定义函数 aggfc，用于接收 rfm[["R", 'F', 'M']].apply(lambda x: x - x.mean())
的返回值
def aggfc(k):
    # 根据"R", 'F', 'M'的正负值对用户进行分类:正数用 1 表示，负数用 0 表示
    level = k.apply(lambda x:'1' if x >0 else '0')
    leable = level.R+level.M+level.F
    # 根据正负数编码,将用户分为 8 个类型
    d = {
        '111': '1 类客户',
        '011': '2 类客户',
        '101': '3 类客户',
        '001': '4 类客户',
```

```
            '110': '5 类客户',
            '010': '6 类客户',
            '100': '7 类客户',
            '000': '8 类客户'
        }
        result = d[leable]
        return result
    # 使用自定义函数 aggfc 对用户进行分类，并赋值给新的字段 label
    df_rfm['label'] = df_rfm[["R",'F',"M"]].apply(lambda x:x-x.mean()).
apply(aggfc,axis=1)
    # 根据 label 字段进行分组，并求和
    print(df_rfm.groupby('label').sum())
    # 根据 label 字段进行分组，并统计个数
    print(df_rfm.groupby('label').count())
    # 根据 lable 标签查找 2 类客户数据，并用 g 绿色标注
    df_rfm.loc[df_rfm.label == '2 类客户','color'] = 'g'
    # 根据 lable 标签查找非 2 类客户数据，并用 r 红色标注
    df_rfm.loc[~(df_rfm.label == '2 类客户'),'color'] = 'r'
    # 使用 scatter 将 F 和 R 作为参数，颜色使用前面的 g 和 r
    df_rfm.plot.scatter("F",'R',c = df_rfm.color)
    plt.show()
    exit()
```

输出如图 4-37 和图 4-38 所示。从图 4-37 中可以看出，2 类客户人数最多，有 4176，并且 M 字段最高，即消费总金额最高，是重要的销售对象。从图 4-38 中可以看出，虽然 2 类客户的单笔消费都比较低且时间间隔较高，但是其总量很大。

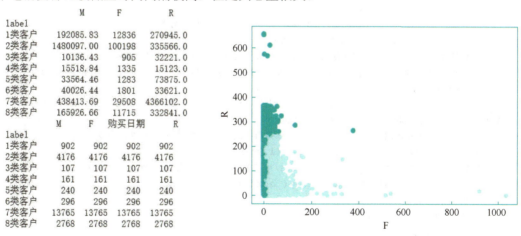

图 4-37　8 类用户的求和与个数统计　　　图 4-38　2 类客户和非 2 类客户的分布情况

13）对用户进行分类之后，现在需要根据每个用户的第一次和最后一次消费时间进行分析，发现其中的新客户、回流客户（再次消费的客户）、活跃客户和不活跃客户。

```
    # 设置自定义时间字段用于索引
    cols = ['2020-01-01', '2020-02-01', '2020-03-01', '2020-04-01',
            '2020-05-01', '2020-06-01', '2020-07-01', '2020-08-01',
            '2020-09-01', '2020-10-01', '2020-11-01', '2020-12-01',
            '2021-01-01', '2021-02-01', '2021-03-01', '2021-04-01',
            '2021-05-01', '2021-06-01']
```

```python
# 自定义函数 active_status 用于区分活跃、不活跃、回流和新用户这四类用户
# 这里区分四类用户的标准是根据前面每个用户第一次和最后一次消费时间
def active_status(data):
    status = []
    # 因为上面的 col 自定义了 18 个字段,所以每行有 18 个数据,因此这里要循环 18 次
    for i in range(18):

        #若本月没有消费
        if data[i] == 0:
            if len(status) > 0:
                if status[i-1] == 'unreg': # unreg 未注册用户
                    status.append('unreg')
                else:
                    status.append('unactive') # 不活跃用户
            else:
                status.append('unreg')

        #若本月有消费
        else:
            if len(status) == 0:
                status.append('new')
            else:
                if status[i-1] == 'unactive':
                    status.append('return')# 回流用户
                elif status[i-1] == 'unreg':
                    status.append('new')# 新用户
                else:
                    status.append('active') # 活跃用户
    return pd.Series(status,index = cols)
# 使用 pivot_table 透视表功能,将 df 数据设置为 index 索引:顾客 ID
# columns 列细分通用名:月份,values 列:购买日期,aggfunc 聚合函数为 count,并使用 fillna 填充 na 为 0
pivot_count = df_data.pivot_table(index = '顾客 ID', columns = '月份', values = '购买日期', aggfunc = 'count').fillna(0)
# 使用 applymap(lambda x:1 if x>0 else 0)根据统计结果进行分类
df_spend=pivot_count.applymap(lambda x:1 if x>0 else 0)
# 使用 apply(active_status,axis = 1)按行传递给自定义函数 active_status
pivot_statu = df_spend.apply( active_status,axis = 1)
# 输出前 5 行
pivot_statu.head()
```

输出如图 4-39 所示。

顾客ID	2020-01-01	2020-02-01	2020-03-01	2020-04-01	2020-05-01	2020-06-01	2020-07-01	2020-08-01	2020-09-01	2020-10-01	2020-11-01	2020-12-01	2021-01-01	2021-02-01	2021-03-01	2021-04-01	2021-05-01
1	unreg	unreg	unreg	unreg	unreg	unreg	new	unactive	unactive	unactive	unactive	unactive	unactive	unactive	unactive	unactive	u
2	unreg	unreg	unreg	unreg	unreg	unreg	new	unactive	unactive	unactive	unactive	unactive	unactive	unactive	unactive	unactive	u
3	unreg	unreg	unreg	unreg	new	unactive	return	unactive	return	unactive	unactive	unactive	unactive	unactive	unactive	unactive	u
4	unreg	unreg	unreg	unreg	unreg	unreg	new	unactive	unactive	unactive	unactive	unactive	unactive	unactive	return	unactive	unactive
5	new	unactive	unactive	unactive	unactive	unactive	return	unactive	unactive	unactive	unactive	unactive	unactive	unactive	unactive	unactive	unactive

图 4-39 对客户进行分类

```
# 使用 replace('unreg',np.NAN)将 np.NAN 替换为 unreg,并统计个数
print(pivot_statu.replace('unreg',np.NAN).apply(lambda
x:pd.value_counts(x)))
```

输出如图 4-40 所示。

```
            2020-01-01  2020-02-01  2020-03-01  2020-04-01  2020-05-01
active             NaN       448.0         541         497         443
new             1460.0      1016.0        1154         577         447
return             NaN         NaN         271         300         525
unactive           NaN      1012.0        1664        2833        3239

            2020-06-01  2020-07-01  2020-08-01  2020-09-01  2020-10-01
active             425         460        1097        1646      1661.0
new                453        5711        6474        5122         NaN
return             551        1161        1920        2108      1006.0
unactive          3678        3486        7801       13538     19747.0

            2020-11-01  2020-12-01  2021-01-01  2021-02-01  2021-03-01
active           806.0       699.0       706.0         577       498.0
new                NaN         NaN         NaN           1         NaN
return          1276.0      1516.0      1365.0        1100      1147.0
unactive       20332.0     20199.0     20343.0       20737     20770.0

            2021-04-01  2021-05-01  2021-06-01
active           507.0       601.0       599.0
new                NaN         NaN         NaN
return          1244.0      1329.0      1170.0
unactive       20664.0     20485.0     20646.0
```

图 4-40 np.NAN 替换为 unreg 后的统计结果

14) 统计每月活跃、新用户、回流用户和不活跃用户的占比。

```
# 统计每月活跃、新用户、回流用户和不活跃用户的占比
pivot_statu = df_spend.apply(active_status, axis=1)
new_point = pivot_statu.replace('unreg',np.NAN).apply(lambda x:pd.value_
counts(x))
# 按行 axis = 1 统计每月活跃、新用户、回流用户和不活跃用户的占比 x:x/x.sum()、使用
fillna 填充缺失值为 0
print(new_point.fillna(0).T.apply(lambda x:x/x.sum(),axis = 1))
```

输出如下所示。

```
              active       new      return     nactive
2020-01-01  0.000000  1.000000  0.000000  0.000000
2020-02-01  0.180937  0.410339  0.000000  0.408724
2020-03-01  0.149036  0.317906  0.074656  0.458402
2020-04-01  0.118136  0.137152  0.071310  0.673401
2020-05-01  0.095187  0.096046  0.112806  0.695960
2020-06-01  0.083219  0.088702  0.107891  0.720188
2020-07-01  0.042522  0.527916  0.107321  0.322241
2020-08-01  0.063440  0.374393  0.111034  0.451133
2020-09-01  0.073436  0.228518  0.094048  0.603998
2020-10-01  0.074105  0.000000  0.044883  0.881012
2020-11-01  0.035960  0.000000  0.056929  0.907112
2020-12-01  0.031186  0.000000  0.067636  0.901178
2021-01-01  0.031498  0.000000  0.060899  0.907602
2021-02-01  0.025742  0.000045  0.049074  0.925139
2021-03-01  0.022217  0.000000  0.051171  0.926612
2021-04-01  0.022619  0.000000  0.055499  0.921883
```

```
2021-05-01  0.026812  0.000000  0.059291  0.913897
2021-06-01  0.026723  0.000000  0.052197  0.921080
```

15）统计回流用户数据。

```
# 使用 pivot_table 函数透视表，index 索引为用户 ID，values 列为订购数量
# columns 列细分通用名为月份，aggfunc 聚合函数为 mean，fillna 填充缺失值为 0
pivot_amount = df_data.pivot_table(index = '顾客 ID',columns = '月份',
values = '消费金额',aggfunc = 'mean').fillna(0)
# 根据订购金额的多少判断为 1 或 0，0 代表当月消费过，次月没有消费，1 代表当月消费过，次月
依然消费
pivot_spend = pivot_amount.applymap(lambda x:1 if x>0 else 0)
# 根据月份字段进行排序，并转换类型为字符串，再去重
col_month = df_data.月份.sort_values().astype('str').unique()
# 给 pivoted_amount 的列细分通用名修改为 columns_month
pivot_amount.columns = col_month
# 自定义函数 purchase_return 用于统计回流用户
def purchase_return(data):
    status = []
    for i in range(17):
        if data[i] == 1:
            if data[i+1] ==1:
                status.append(1)
            if data[i+1] == 0:
                status.append(0)
        else:
            status.append(np.NaN)
    status.append(np.NaN)
    return pd.Series(status,index=cols)
# 根据前面统计的订购金额情况（0 或 1），输入到函数 purchase_return 中，得出回流用户数据
pivot_spend_return = pivot_spend.apply(purchase_return,axis = 1)
# 老客户的回购率计算
(pivot_spend_return.sum()/pivot_spend_return.count()).plot()
plt.xticks(rotation = 45)  # 横坐标的数据旋转 45°
plt.show()
```

输出如图 4-41 所示。

任务 3　使用 Pandas 实现电商销售数据分析

使用 Pandas 实现电商销售数据分析，包括获取和查看数据概况、数据清洗、数据可视化。

1）获取和查看数据概况，包括数据基本信息、数据前几行样例信息、数据维度、数据空值统计。该源数据含有 24 个字段，44703 行数据，数据类型包括 5 个 float64 型，2 个 int64 和 17 个 object 型。数据中部分字段存在空值情况。

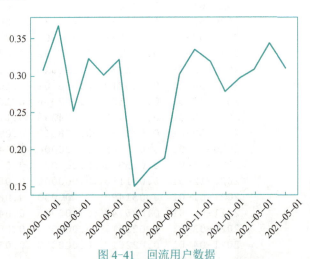

图 4-41　回流用户数据

```
# 查看数据
import pandas as pd
# 使用 read_csv 函数读取文件
```

```python
df_data= pd.read_csv('./dataset.csv',encoding='utf-8')
# 查看数据前 5 行
# df_data.head()
```

输出如图 4-42 所示，为数据前 5 行。

序号	订单号	下单日期	发货时间	发货模式	顾客ID	顾客姓名	顾客类型	所在城市	所在州	...	产品ID	产品类别	产品子类别	产品名	购买价格	购买数量	优惠
0	1	IN-2011-47883	2017/1/1	2017/1/8	Standard Class	JH-15985	Joseph Holt	Consumer	Wagga Wagga	New South Wales	...	OFF-3U-10000618	Office Supplies	Supplies	Acme Trimmer, High Speed	120.366	3	0.1
1	2	IN-2011-47883	2017/1/1	2017/1/8	Standard Class	JH-15985	Joseph Holt	Consumer	Wagga Wagga	New South Wales	...	OFF-PA-10001968	Office Supplies	Paper	Eaton Computer Printout Paper, 8.5 x 11	55.242	2	0.1
2	3	IN-2011-47883	2017/1/1	2017/1/8	Standard Class	JH-15985	Joseph Holt	Consumer	Wagga Wagga	New South Wales	...	FUR-FU-10003447	Furniture	Furnishings	Eldon Light Bulb, Duo Pack	113.670	5	0.1
3	4	IT-2011-3647632	2017/1/1	2017/1/5	Second Class	EM-14140	Eugene Moren	Home Office	Stockholm	Stockholm	...	OFF-PA-10001492	Office Supplies	Paper	Enermax Note Cards, Premium	44.865	3	0.5
4	5	HU-2011-1220	2017/1/1	2017/1/5	Second Class	AT-735	Annie Thurman	Consumer	Budapest	Budapest	...	OFF-TEN-10001585	Office Supplies	Storage	Tenex Box, Single Width	66.120	4	0.0

5 rows × 24 columns

图 4-42　查看数据前 5 行

```
# 查看数据基本信息
df_data.info()
```

输出如图 4-43 所示。

```
<class 'pandas.core.frame.DataFrame'>
RangeIndex: 44703 entries, 0 to 44702
Data columns (total 24 columns):
 #   Column      Non-Null Count  Dtype
---  ------      --------------  -----
 0   序号          44703 non-null  int64
 1   订单号         44703 non-null  object
 2   下单日期        44703 non-null  object
 3   发货时间        44703 non-null  object
 4   发货模式        44694 non-null  object
 5   顾客ID        44703 non-null  object
 6   顾客姓名        44703 non-null  object
 7   顾客类型        44703 non-null  object
 8   所在城市        44703 non-null  object
 9   所在州         44703 non-null  object
 10  所在国家        44703 non-null  object
 11  邮政编码         8677 non-null  float64
 12  交易市场        44703 non-null  object
 13  交易地区        44703 non-null  object
 14  产品ID        44703 non-null  object
 15  产品类别        44703 non-null  object
 16  产品子类别       44703 non-null  object
 17  产品名         44703 non-null  object
 18  购买价格        44703 non-null  float64
 19  购买数量        44703 non-null  int64
 20  优惠          44703 non-null  float64
 21  利润          44703 non-null  float64
 22  邮费          44703 non-null  float64
 23  订单优先级       44703 non-null  object
dtypes: float64(5), int64(2), object(17)
memory usage: 8.2+ MB
```

图 4-43　查看数据基本信息

```
# 查看数据维度
df_data.shape
```

输出如下：

```
(51101, 24)
# 查看描述统计信息
df_data.describe()
```

输出如图4-44所示。

	序号	邮政编码	购买价格	购买数量	优惠	利润	邮费
count	44703.000000	8677.000000	44703.000000	44703.000000	44703.000000	44703.000000	44703.000000
mean	22440.567456	55135.409358	246.407723	3.464958	0.142780	28.598723	26.351124
std	12998.733801	32024.819260	487.205651	2.265946	0.213371	172.496308	57.336587
min	1.000000	1040.000000	0.444000	1.000000	0.000000	-6599.978000	0.000000
25%	11175.500000	23223.000000	30.744000	2.000000	0.000000	0.000000	2.620000
50%	22350.000000	57103.000000	85.230000	3.000000	0.000000	9.288000	7.810000
75%	33751.500000	90004.000000	250.650000	5.000000	0.200000	36.810000	24.380000
max	50827.000000	99301.000000	22638.480000	14.000000	1.500000	8399.976000	933.570000

图4-44 查看描述统计信息

```
# 非空统计和空值数量
df_data.count()
df_data.isnull().sum()
```

输出如图4-45和图4-46所示。

序号	44703		序号	0
订单号	44703		订单号	0
下单日期	44703		下单日期	0
发货时间	44703		发货时间	0
发货模式	44694		发货模式	9
顾客ID	44703		顾客ID	0
顾客姓名	44703		顾客姓名	0
顾客类型	44703		顾客类型	0
所在城市	44703		所在城市	0
所在州	44703		所在州	0
所在国家	44703		所在国家	0
邮政编码	8677		邮政编码	36026
交易市场	44703		交易市场	0
交易地区	44703		交易地区	0
产品ID	44703		产品ID	0
产品类别	44703		产品类别	0
产品子类别	44703		产品子类别	0
产品名	44703		产品名	0
购买价格	44703		购买价格	0

图4-45 非空统计　　　　　　图4-46 统计数据空值数量

2）数据清洗，包括时间数据类型转、数据无效值过滤、数据去重、数据空值填充。

```
# 清洗数据
# 将其转化成时间（发货时间，下单日期均为object，需要先转化成时间）
df_data['发货时间'] = pd.to_datetime(df_data['发货时间'])
df_data['下单日期'] = pd.to_datetime(df_data['下单日期'])
# 查看数据基本信息
df_data.info()
```

输出如图4-47所示。

```
#找出发货时间早于下单日期的记录
df_data[df_data['发货时间']<df_data['下单日期']]
#删去发货时间早于下单日期的记录,且在原数据上进行修改
df_data.drop(index=df_data[df_data['发货时间']<df_data['下单日期']].index, inplace=True)
df_data
```

```
<class 'pandas.core.frame.DataFrame'>
RangeIndex: 44703 entries, 0 to 44702
Data columns (total 24 columns):
 #   Column   Non-Null  Count     Dtype
---  ------   --------  ------    -----
 0   序号      44703    non-null       int64
 1   订单号    44703    non-null       object
 2   下单日期   44703    non-null  datetime64[ns]
 3   发货时间   44703    non-null  datetime64[ns]
 4   发货模式   44694    non-null       object
 5   顾客ID    44703    non-null       object
 6   顾客姓名   44703    non-null       object
 7   顾客类型   44703    non-null       object
 8   所在城市   44703    non-null       object
 9   所在州    44703    non-null       object
 10  所在国家   44703    non-null       object
 11  邮政编码    8677    non-null      float64
 12  交易市场   44703    non-null       object
 13  交易地区   44703    non-null       object
 14  产品ID    44703    non-null       object
 15  产品类别   44703    non-null       object
 16  产品子类别  44703    non-null       object
 17  产品名    44703    non-null       object
 18  购买价格   44703    non-null      float64
 19  购买数量   44703    non-null        int64
 20  优惠     44703    non-null      float64
 21  利润     44703    non-null      float64
 22  邮费     44703    non-null      float64
 23  订单优先级  44703    non-null       object
dtypes: datetime64[ns](2), float64(5), int64(2), object(15)
memory usage: 8.2+ MB
```

图 4-47　时间类型转换

输出如图 4-48 所示。

	序号	订单号	下单日期	发货时间	发货模式	顾客ID	顾客姓名	顾客类型	所在城市	所在州	...	产品ID	产品类别	产品子类别	产品名	购买
0	1	IN-2011-47883	2017-01-01	2017-01-08	Standard Class	JH-15985	Joseph Holt	Consumer	Wagga Wagga	New South Wales	...	OFF-SU-10000618	Office Supplies	Supplies	Acme Trimmer, High Speed	120.36
1	2	IN-2011-47883	2017-01-01	2017-01-08	Standard Class	JH-15985	Joseph Holt	Consumer	Wagga Wagga	New South Wales	...	OFF-PA-10001968	Office Supplies	Paper	Eaton Computer Printout Paper, 8.5 x 11	55.24
2	3	IN-2011-47883	2017-01-01	2017-01-08	Standard Class	JH-15985	Joseph Holt	Consumer	Wagga Wagga	New South Wales	...	FUR-FU-10003447	Furniture	Furnishings	Eldon Light Bulb, Duo Pack	113.67
3	4	IT-2011-3647632	2017-01-01	2017-01-05	Second Class	EM-14140	Eugene Moren	Home Office	Stockholm	Stockholm	...	OFF-PA-10001492	Office Supplies	Paper	Enermax Note Cards, Premium	44.86
4	5	HU-2011-1220	2017-01-01	2017-01-05	Second Class	AT-735	Annie Thurman	Consumer	Budapest	Budapest	...	OFF-TEN-10001585	Office Supplies	Storage	Tenex Box, Single Width	66.12
⋮	⋮	⋮	⋮	⋮	⋮	⋮	⋮	⋮	⋮	⋮		⋮	⋮	⋮	⋮	
44698	44923	CG-2014-4740	2020-09-26	2020-09-30	Standard Class	MH-7440	Mark Haberlin	Corporate	Likasi	Katanga	...	FUR-DAN-10004745	Furniture	Bookcases	Dania Corner Shelving, Traditional	244.14
44699	44924	MX-2014-144806	2020-09-26	2020-09-28	Second Class	DM-13015	Darrin Martin	Consumer	Quetzaltenango	Quezaltenango	...	OFF-PA-10001685	Office Supplies	Paper	Eaton Parchment Paper, Recycled	80.28
44700	44925	GH-2014-8440	2020-09-26	2020-10-01	Standard Class	SR-10740	Steven Roelle	Home Office	Kumasi	Ashanti	...	OFF-ELD-10003181	Office Supplies	Storage	Eldon Box, Industrial	42.96
44701	44926	MX-2014-114818	2020-09-26	2020-09-28	First Class	RS-19870	Roy Skaria	Home Office	San Miguelito	Panama	...	FUR-FU-10002422	Furniture	Furnishings	Eldon Photo Frame, Duo Pack	69.80
44702	50827	CA-2014-136238	2020-12-27	2021-01-02	Standard Class	KB-16240	Karen Bern	Corporate	Odessa	Texas	...	OFF-PA-10004285	Office Supplies	Paper	Xerox 1959	16.03

44689 rows × 24 columns

图 4-48　过滤掉发货时间早于下单日期的数据

```
# 根据销售额字段检查是否存在销售额小于零的数据，如果有则要删除。
df_data[df_data.购买价格<0]
```

输出如图 4-49 所示。

序号	订单号	下单日期	发货时间	发货模式	顾客ID	顾客姓名	顾客类型	所在城市	所在州	...	产品ID	产品类别	产品子类别	产品名	购买价格	购买数量	优惠	利润	邮费	订单优先级

0 rows × 24 columns

图 4-49 销售额小于零的数据

```
#序号不重复的个数    （经过上面的数据处理后，由 data.shape#(51097, 24)知序号不重复的个数为 51094）
df_data.序号.unique().size
#取出重复的记录
df_data[df_data.序号.duplicated()]
#删去序号重复的记录，且在原数据上进行修改
df_data.drop(index=df_data[df_data.序号.duplicated()].index, inplace=True)
# 查看发货模式空值
print(df_data[df_data.发货模式.isnull()]['发货模式'])
```

输出如图 4-50 所示。

```
# 对空值进行修补
# 从选择的某个轴返回这个众数，如果缺失就用 NaN 填充，返回 dateframe 类型
print(df_data.发货模式.mode()[0])
```

输出如下：

```
Standard Class
# 使用 Standard Class 进行空值填充
df_data['发货模式'].fillna(value=df_data.发货模式.mode()[0], inplace=True)
df_data.info()
```

输出如图 4-51 所示。

```
13666    NaN
13674    NaN
18385    NaN
32940    NaN
33392    NaN
33395    NaN
37937    NaN
44016    NaN
44094    NaN
Name: 发货模式, dtype: object
```

图 4-50 查看发货模式空值

```
Standard Class
<class 'pandas.core.frame.DataFrame'>
Int64Index: 44686 entries, 0 to 44702
Data columns (total 24 columns):
 #   Column      Non-Null Count  Dtype
 0   序号          44686 non-null  int64
 1   订单号         44686 non-null  object
 2   下单日期        44686 non-null  datetime64[ns]
 3   发货时间        44686 non-null  datetime64[ns]
 4   发货模式        44686 non-null  object
 5   顾客ID        44686 non-null  object
 6   顾客姓名        44686 non-null  object
 7   顾客类型        44686 non-null  object
 8   所在城市        44686 non-null  object
 9   所在州         44686 non-null  object
 10  所在国家        44686 non-null  object
 11  邮政编码        8674 non-null   float64
 12  交易市场        44686 non-null  object
 13  交易地区        44686 non-null  object
 14  产品ID        44686 non-null  object
 15  产品类别        44686 non-null  object
 16  产品子类别       44686 non-null  object
 17  产品名         44686 non-null  object
 18  购买价格        44686 non-null  float64
 19  购买数量        44686 non-null  int64
 20  优惠          44686 non-null  float64
 21  利润          44686 non-null  float64
 22  邮费          44686 non-null  float64
 23  订单优先级       44686 non-null  object
dtypes: datetime64[ns](2), float64(5), int64(2), object(15)
memory usage: 9.5+ MB
```

图 4-51 发货模式字段填充后的结果

3）分别统计每年的销售总金额、利润总金额、总邮寄费用和销售总数。

```
#分别取订单日期的年、月、季数据
df_data['下单年份'] = df_data['下单日期'].dt.year
df_data ['下单月份'] = df_data ['下单日期'].dt.month
df_data ['下单季度'] = df_data ['下单日期'].dt.to_period('Q')
result = df_data [['下单日期','下单年份','下单月份', '下单季度']].head()
```

```python
# print(result)
# 获取每年的销售总金额
year_sale = df_data.groupby(by='下单年份')['购买价格'].sum()
print(year_sale)
# 获取每年的利润总金额
sale_pf = df_data.groupby(by='下单年份')['利润'].sum()
print(sale_pf)
# 获取每年的总邮寄费用
sale_shipcost = df_data.groupby(by='下单年份')['邮费'].sum()
print(sale_shipcost)
# 获取每年的销售总数
sale_qt = df_data.groupby(by='下单年份')['购买数量'].sum()
print(sale_qt)
```

输出如图 4-52 所示。

4) 以折线图呈现每年销售总金额变化，饼图呈现每年利润总金额变化，柱状图呈现每年邮费总金额变化。从中可以发现，排除 2020 年的数据，每年销售总金额、每年利润总金额和每年邮费总金额都是逐年递增的，说明业务持续向好。

```python
import matplotlib.pyplot as plt
import matplotlib as mpl
# 设置字体
mpl.rcParams['font.sans-serif'] = ['SimHei']
# 设置风格
plt.style.use('ggplot')
# 以折线图呈现每年销售总金额变化
year_sale = df_data.groupby(by='下单年份')['购买价格'].sum()
plt.plot(year_sale.index, year_sale)
```

输出如图 4-53 所示。

```
下单年份
2017    2.254364e+06
2018    2.664357e+06
2019    3.342698e+06
2020    2.751189e+06
Name: 购买价格, dtype: float64
下单年份
2017    247367.76678
2018    304619.06114
2019    399497.16564
2020    326520.72970
Name: 利润, dtype: float64
下单年份
2017    243939.14
2018    282033.28
2019    357773.47
2020    294064.57
Name: 邮费, dtype: float64
下单年份
2017    31342
2018    37869
2019    47146
2020    38490
Name: 购买数量, dtype: int64
```

图 4-52 获取每年销售总金额、利润总金额、总邮费和购买总数量

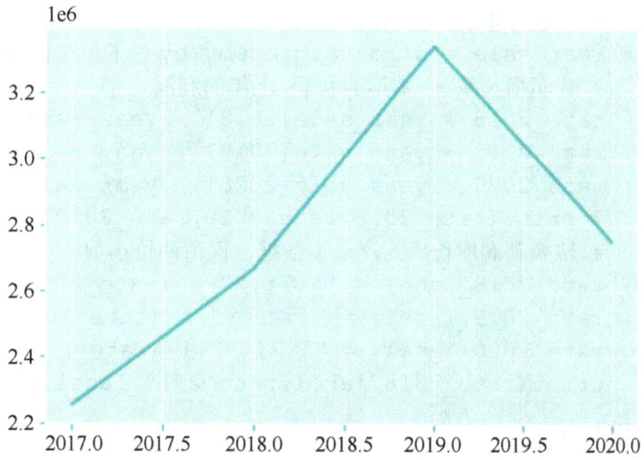

图 4-53 每年销售总金额变化

```python
# 以饼图呈现每年利润总金额变化
sale_pf = df_data.groupby(by='下单年份')['利润'].sum()
explode = (0, 0.1, 0, 0)
```

```
labels = sale_pf.index
colors = ['yellowgreen', 'gold', 'lightskyblue', 'lightcoral']
plt.pie(sale_pf, explode=explode, labels=labels, colors=colors, autopct=
'%1.1f%%', shadow=True,startangle=90)
```

输出如图 4-54 所示。

```
# 以柱状图呈现每年邮费总金额变化
sale_shipcost = df_data.groupby(by='下单年份')['邮费'].sum()
plt.bar(sale_shipcost.index, sale_shipcost, align = 'center',color = 'b')
```

输出如图 4-55 所示。

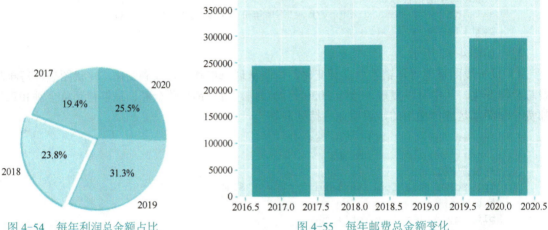

图 4-54　每年利润总金额占比　　　　　　图 4-55　每年邮费总金额变化

5）统计并以可视化组合图（柱状图和折线图）方式呈现每年利润的增长率。

```
#分别取订单日期的年、月、季数据
df_data['下单年份'] = df_data ['下单日期'].dt.year
df_data ['下单月份'] = df_data ['下单日期'].dt.month
df_data ['下单季度'] = df_data ['下单日期'].dt.to_period('Q')
result = df_data [['下单日期','下单年份','下单月份', '下单季度']].head()
# print(result)
# 获取每年的销售利润总金额
year_sale = df_data.groupby(by='下单年份')['利润'].sum()
# 利润增长率 = 本年的利润/上年的利润 - 1
rate_2018 = year_sale[2018] / year_sale[2017] - 1
rate_2019 = year_sale[2019] / year_sale[2018] - 1
rate_2020 = year_sale[2020] / year_sale[2019] - 1
# print(rate_2018,rate_2019,rate_2020)
# 转换利润增长率为浮点型数据，保留两位小数
rate_2018_label = "%.2f%%" % (rate_2018 * 100)
rate_2019_label = "%.2f%%" % (rate_2019 * 100)
rate_2020_label = "%.2f%%" % (rate_2020 * 100)
print(rate_2018_label,rate_2019_label,rate_2020_label)
```

输出如下：

```
23.14% 31.15% -18.27%
```

```
# 使用 DataFrame 创建 DataFrame 对象，并传入指定的列：sales_profit, sales_rate 和
sales_rate_label, 及其对应的值
sale_rate = pd.DataFrame(
    {'sale_pf':year_sale,
```

```
            'sale_rate':[0,rate_2018,rate_2019,rate_2020],
'rate_label':['0.00%',rate_2018_label,rate_2019_label,rate_2020_label]
            })
    # 给可视化组合图准备数据，以 y1 和 y2 分别给柱状图和折线图的 y 轴赋值
    y1 = sale_rate['sale_pf']
    y2 = sale_rate['sale_rate']
    # 通过 for 循环获得 sales_rate.index 的年份
    x = [str(value) for value in sale_rate.index.tolist()]
    # 新建 figure 对象
    fig=plt.figure()
    # 新建子图 1
    ax1=fig.add_subplot(1,1,1)

    # ax2 与 ax1 共享 x 轴
    ax2 = ax1.twinx()
    # 绘制柱状图
    ax1.bar(x,y1,color = 'r')
    # 绘制折线图
    ax2.plot(x,y2,marker='*',color = 'b')
    # 设置柱状图和折线图标签和标题
    ax1.set_xlabel('年份')
    ax1.set_ylabel('净值')
    ax2.set_ylabel('增长率')
    ax1.set_title('净值与增长率')
```

plt.show()输出如图 4-56 所示。

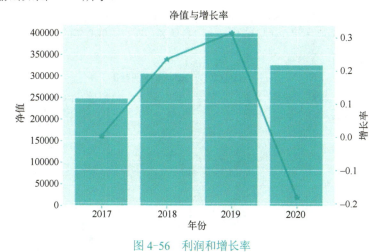

图 4-56　利润和增长率

6）统计并以可视化饼图方式呈现不同交易地区的邮费总额占比。

```
#不同交易地区的邮费总额
region = df_data.groupby(by='交易地区')['邮费'].sum()
region.plot(kind='pie')
region.plot(kind='pie',autopct="%1.2f%%",title='不同区域的邮费价格')
```

输出如图 4-57 所示。

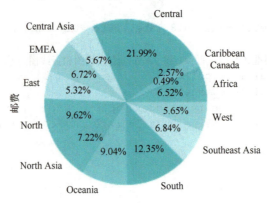

图 4-57　不同交易地区的邮费总额占比

7) 统计并以可视化柱状图方式呈现各地区每一年的销售数量。

```
# 各地区每一年的销售数量
area = df_data.groupby(by=['交易地区','下单年份'],as_index=False)['购买数量'].sum()
# sales_area

# 使用数据透视表重新整理数据
area = pd.pivot_table(area,
                      index='交易地区',
                      columns='下单年份',
                      values='购买数量')
area
# 绘制图形
colors = ['yellowgreen', 'gold', 'lightskyblue', 'lightcoral']
area.plot(kind = 'bar',title = '各地区2017～2020年的销售数量',color=colors)
```

输出如图 4-58 所示。

图 4-58　各地区 2017～2020 年的销售数量

8) 统计 2018 年的客户成分数据，根据'顾客 ID'、'下单日期'、'购买价格'和'订单'字段，对用户进行类别分析。具体方式：以'顾客 ID'字段为索引，分别计算'下单日期'、'购买价格'和'订单'字段与各自的平均值相减的结果（正数表示 1，负数表示 0），并进行对应的编码标识之后，对不同的客户进行分类，实现未来的精准营销。

```
#取出订单日期为 2018 年的数据
data_2018 = df_data [df_data ['下单年份']==2018]
#取出三列
data_2018 = data_2018[['顾客 ID','下单日期','购买价格']]
#重新复制给 customdf
df_custom = data_2018.copy()
#设置 customdf 的索引为顾客 ID, drop=True 就是把原来的索引 index 列去掉, 重置 index,
inplace=True 表示直接在原数组上对数据进行修改
df_custom.set_index('顾客 ID',drop=True,inplace=True)
#增加一个辅助列，每一条订单都计数为 1
df_custom['订单'] = 1
#customdf
#透视图 index 为索引, values 为列名, aggfunc 为操作函数
df_rfm = df_custom.pivot_table(index=['顾客 ID'],
                  values=['下单日期','订单','购买价格'],
                  aggfunc={'下单日期':'max',
                           '订单':'sum',
                           '购买价格':'sum'})
# 计算用户最后一次购买时间与第一次购买时间间隔，并添加到新增列 R
df_rfm['R'] = (df_rfm.下单日期.max()-df_rfm.下单日期).dt.days
# 重命名'购买价格':'M','订单':'F'
df_rfm.rename(columns={'购买价格':'M','订单':'F'},inplace=True)
# 自定义函数 rfm_func 用于给不同的客户进行编码。分别计算'下单日期'、'购买价格'和'订单'
字段与各自的平均值相减的结果（正数表示 1，负数表示 0）
def rfm_func(x):
    level = x.apply(lambda x: "1" if x >= 0 else '0')
    label = level.R + level.F + level.M
    d = {
        '011':'重要价值客户',
        '111':'重要唤回客户',
        '001':'重要深耕客户',
        '101':'重要挽留客户',
        '010':'潜力客户',
        '110':'一般维持客户',
        '000':'新客户',
        '100':'流失客户'
    }
    result = d[label]
    return result
# 使用'R'、'F'、'M'与各自的平均值相减的方法，判断其客户重要程度
result1= df_rfm[['R','F','M']].apply(lambda x:x-x.mean())
# 根据'R'、'F'、'M'与各自的平均值相减的结果，对不同的客户进行分类
result2 = result1.apply(rfm_func,axis=1)
result2
# 将客户分类的结果添加到 rfmdf 中，增加新的列：labels
df_rfm['labels'] = result2
df_rfm
```

输出如图 4-59 所示。

顾客ID	下单日期	F	M	R	labels
AA-10315	2018-10-04	2	2740.37000	88	重要挽留客户
AA-10375	2018-12-03	8	1756.88700	28	潜力客户
AA-10480	2018-12-28	11	4941.57978	3	重要价值客户
AA-10645	2018-12-14	20	4147.38116	17	重要价值客户
AA-375	2018-09-27	4	114.61200	95	流失客户
...	...	...	...	...	...
YS-11880	2018-11-27	4	1102.02000	34	新客户
YS-21880	2018-12-08	23	6497.41200	23	重要价值客户
ZC-21910	2018-10-04	18	4714.74480	88	重要唤回客户
ZD-11925	2018-08-10	2	172.62600	143	流失客户
ZD-21925	2018-12-12	11	3188.29750	19	重要价值客户

1372 rows × 5 columns

图 4-59 不同客户的分类

练习题

使用 Pandas 实现某地区电影票房数据分析。
1）获取指定的 csv 源数据。
2）获取计算指定电影《简单爱》的上映天数及日均票房。
3）将结果保存到 movie_data.dat 文件中。

项目 5　Matplotlib 数据分析可视化库

【项目分析】

本项目旨在理解和掌握 Matplotlib 数据分析工具的基本概念和基础语法。具体内容如下。

1）Matplotlib 常见的数据分析图形：散点图、条形图、折线图、饼图、直方图、箱形图和组合图以及实例。

2）Matplotlib 数据可视化分析案例介绍。

【知识准备】

5.1　Matplotlib 的基本概念

5.1.1　Matplotlib 基础理论和引用方法

"数据可视化"可以帮助用户理解数据，通过各种图形能够非常形象、生动地描述数据和业务背后的故事。

Matplotlib 是一个综合库，用于在 Python 中创建静态、动画和交互式可视化。Matplotlib 中丰富的 API 可以绘制各种图形，包括散点图、条形图、折线图、饼图、直方图、箱形图以及各种组合图，与前面 NumPy 和 Pandas 的引用方式类似，这里通过引用 Anaconda 自带的 Python 解释器就可以导入 Pandas 包，如图 5-1 所示。

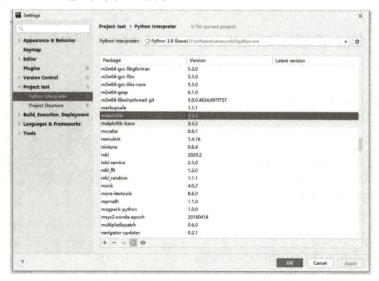

图 5-1　引用 Matplotlib

5.1.2 散点图

散点图是通过横坐标和纵坐标表示两个变量之间分布形态的一种关系图形，能够直观地呈现出两个变量之间相互影响的总体趋势。

散点图 scatter 函数的定义是：

```
scatter(x, y, s=None, c=None, marker=None, cmap=None, norm=None, vmin=None,
vmax=None, alpha=None, linewidths=None, edgecolors=None, hold=None, data=None,
**kwargs)
```

参数详解见表 5-1。

表 5-1　散点图 scatter 函数参数详解

常用参数名称	功能描述
x	表示 x 轴数据的向量，长度与 y 相等，接收 array
y	表示 y 轴数据的向量，长度与 x 相等，接收 array
s	（可选）表示一个或多个标记的大小，接收 array
c	（可选）表示标记的颜色：RGB 三原色或颜色名称 例如，'red'、'r'或[1 0 0]
marker	（可选）表示标记样式，例如，'o'表示圆圈
cmap	（可选）使用 RGB 三原色的色彩盘，可实现颜色的线性操作
norm	（可选）表示数据亮度[0～1]，浮点型数据
vmin/vmax	（可选）表示数据亮度[0～1]，浮点型数据 如果为 None（无），则使用颜色阵列的最小值和最大值。给定范数时，不建议使用 vmin/vmax
alpha	（可选）表示透明度：[0～1]，1 不透明，0 透明
linewidths	（可选）线宽，表示标记边缘的宽度
edgecolors	（可选）表示轮廓颜色，参数与 c 类似

【实例 5-1】 实现一个基本的散点图。

本实例通过导入 NumPy 和 Matplotlib 库，然后使用 randn 函数生成 1000 个随机数作为 x 轴和 y 轴的数据，最后使用 scatter 和 show 函数接收参数并呈现散点图。

```
import numpy as np
import matplotlib.pyplot as plt
N = 1000
# 使用 randn 函数生成 1000 个随机数作为 x 轴的数据
x = np.random.randn(N)
# 使用 randn 函数生成 1000 个随机数作为 y 轴的数据
y = np.random.randn(N)
# 使用 scatter 函数创建散点图
plt.scatter(x, y)
# 使用 show 函数呈现散点图
plt.show()
```

输出如图 5-2 所示。

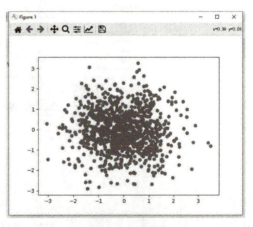

图 5-2　基本的散点图

【实例 5-2】 实现一个自定图标的散点图。

本实例通过导入 NumPy 和 Matplotlib 库，然后使用 seed 函数生成随机数种子，rand 函数生成一组"0～1"均匀分布的随机样本值，x 和 y 各包含 100 个随机值，使用 masked_where 函数作为判断条件，用来显示不同面积大小的图形，使用 scatter 函数接收散点图数据和自定义的图形参数，show 函数用于呈现图形。

```python
# -*- coding: utf-8 -*-
import matplotlib.pyplot as plt
import numpy as np

# 使用 seed 函数设置随机数种子
np.random.seed(19680801)
# 自定义一个生成图形数量的值，这里 100 表示每个图形将生成 100 个图形
N = 100
# 自定义一个阈值用于选择图形面积
r0 = 0.6
# 通过 rand 函数可以返回一组"0～1"均匀分布的随机样本值，x 和 y 各包含 100 个随机值
x = 0.9 * np.random.rand(N)
y = 0.9 * np.random.rand(N)
# 通过 rand 函数可以返回一组"0～1"均匀分布的随机样本值，这里用于计算获得图形面积
area = (20 * np.random.rand(N))**2
c = np.sqrt(area)
r = np.sqrt(x ** 2 + y ** 2)
# 使用 masked_where 函数作为判断条件，选择当前迭代的 area1 和 area2 是否使用 area 的
# 值，用来显示不同面积大小的图形
area1 = np.ma.masked_where(r < r0, area)
area2 = np.ma.masked_where(r >= r0, area)
# 使用 scatter 函数传入 x、y 数据点以及 area1 和 area2 面积，并自定义图形样式为:'^'和'o'
plt.scatter(x, y, s=area1, marker='^', c=c)
plt.scatter(x, y, s=area2, marker='o', c=c)
# 显示散点图
plt.show()
```

输出如图 5-3 所示。

【实例 5-3】实现一个带图例的散点图。

本实例通过导入 NumPy 和 Matplotlib 库，然后使用 seed 函数生成随机数种子，subplots 函数返回一个 figure 图像和子图 ax 的 array 列表，使用 for 循环将列表['tab:blue', 'tab:orange', 'tab:red']作为标签颜色导入，使用 legend 函数显示标签，使用 grid 函数表示是否显示网格，show 函数用于呈现图形。

```python
# -*- coding: utf-8 -*-
import numpy as np
import matplotlib.pyplot as plt
# 使用 seed 函数设置随机数种子
np.random.seed(20210801)
# subplots 函数返回一个 figure 图像和子图 ax 的 array 列表
fig, ax = plt.subplots()
# 使用 for 循环将列表['tab:blue', 'tab:orange', 'tab:red']作为标签颜色导入
for color in ['tab:blue', 'tab:orange', 'tab:red']:
    n = 500  #自定义生成 500 个图像
    x, y = np.random.rand(2, n) #生成 0～1 范围内的随机数各 500 个，分别赋值给 x 和 y
    scale = 100.0 * np.random.rand(n)  # 随机生成图形面积大小
    ax.scatter(x, y, c=color, s=scale, label=color,
               alpha=0.9, edgecolors='none')  #alpha 表示透明度，0 为完全透明，1
# 为不透明；edgecolors 表示是否显示边界颜色
```

```
# 使用 legend 函数显示标签
ax.legend()
# 使用 grid 函数表示是否显示网格
ax.grid(True)
# 显示带图例的散点图
plt.show()
```

输出如图 5-4 所示。

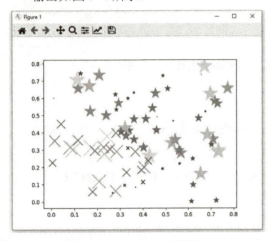

图 5-3 自定义图标的散点图

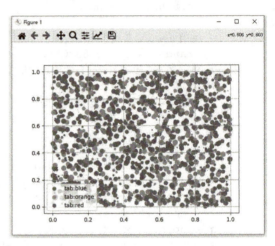

图 5-4 带图例的散点图

5.1.3 条形图

条形图是根据数量的多少画成的长短不同、宽度相同的直条。条形图能够让各个数据的数量大小一目了然，因此比较容易展示数据之间大小的差别。

条形图 bar 函数的定义是：

```
bar (x, height, width = 0.8, bottom = None, *, align = 'center', data = None, ** kwargs)
```

参数详解见表 5-2。

表 5-2 条形图 bar 函数参数详解

常用参数名称	功能描述
x	条形的 x 坐标，接收 array
height	条形的高度，接收 array
width	（可选）条形的宽度，默认值为 0.8
color	（可选）条形图的颜色，例如，"r","b","g","#123465"，默认"b"
bottom	（可选）条形基座的 y 坐标，默认值为 0
align	（可选）条形与 x 坐标的对齐 {'center', 'edge'}，默认值为'center'
linewidth	（可选）边的宽度，0 表示无边框
tick_label	（可选）标签，默认无，使用数字标签

【实例 5-4】 实现一个基本的条形图。

本实例通过导入 Matplotlib 库，自定义条形图 1 和条形图 2 的 x 和 y 轴数据，使用 bar 函数接收 x 和 y 轴数据，使用 title 函数传入标题文字，使用 ylabel 函数传入 y 轴的文字描述，使用 xlabel 函数传入 x 轴的文字描述，show 函数用于呈现图形。

```
from matplotlib import pyplot as plt
#显示中文标签
plt.rcParams['font.sans-serif']=['SimHei'] #显示中文标签
# 条形图 1 的 x 和 y 轴数据
x = [5,8,10]
y = [12,16,6]
# 条形图 2 的 x 和 y 轴数据
x2 = [6,9,11]
y2 = [6,15,7]
# 将条形图 1 的 x 和 y 轴数据传入 bar 函数中，align = 'center'表示图形位置居中排列
plt.bar(x, y, align = 'center')
# 将条形图 2 的 x 和 y 轴数据传入 bar 函数中，color = 'g'表示条形的颜色为绿色，默认是蓝色
plt.bar(x2, y2, color = 'g', align = 'center')
# 使用 title 函数传入标题文字
plt.title('条形图的标题')
# 使用 ylabel 传入 y 轴的文字描述
plt.ylabel('y 轴')
# 使用 xlabel 传入 x 轴的文字描述
plt.xlabel('x 轴')
# 使用 show 函数运行图形
plt.show()
```

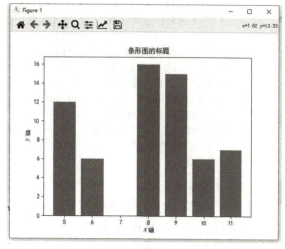

图 5-5　基本条形图

输出如图 5-5 所示。

【实例 5-5】　实现一个水平条形图。

本实例通过导入 NumPy 和 Matplotlib 库，使用 mpl.rcParams 设置字体样式，使用 seed 函数生成随机数种子，使用 subplots 函数返回一个 figure 图像和子图 ax 的 array 列表，使用 rand 函数随机生成 people 元素个数的 0～1 的数字，使用 rand 函数随机生成 people 元素个数的 0～1 的数字，这个 error 用于下面的方差属性 xerr。使用 barh 函数配置水平条形图，使用 set_yticks 函数设置 y 轴标记位置，使用 set_yticklabels 函数设置 y 轴标签内容，使用 invert_yaxis 函数设置 y 轴标签顺序，使用 set_xlabel 函数设置 x 轴标签内容，使用 set_title 函数设置图例标题内容，show 函数用于呈现图形。

```
# -*- coding: utf-8 -*-
import matplotlib.pyplot as plt
import numpy as np
import matplotlib as mpl
# 汉字字体，优先使用楷体，如果找不到楷体，则使用黑体
mpl.rcParams['font.sans-serif'] = ['KaiTi', 'SimHei', 'FangSong']
# 字体大小
mpl.rcParams['font.size'] = 12
# 正常显示负号
mpl.rcParams['axes.unicode_minus'] = False
# 使用 seed 函数设置随机数种子
np.random.seed(20210802)
```

```python
# 使用 subplots() 函数返回一个 figure 图像和子图 ax 的 array 列表
fig, ax = plt.subplots()

# 样例数据
people = ('Simon', 'John', 'Tim', 'Joe', 'Sam')
# 根据 people 的元素个数生成 y 轴的标记个数
y_pos = np.arange(len(people))
# 使用 rand 函数随机生成 people 元素个数的 0~1 的数字
performance = 3 + 10 * np.random.rand(len(people))
# 使用 rand 函数随机生成 people 元素个数的 0~1 的数字,这个 error 用于下面的方差属性 xerr
error = np.random.rand(len(people))
# 使用 barh 函数配置水平条形图。y 轴传入 y_pos,x 轴传入 performance,x 轴方向的数据点误差计算方法传入 error,布局 align 传入 center 居中
ax.barh(y_pos, performance, xerr=error, align='center')
# 将刻度放在数据范围中
ax.set_yticks(y_pos)
# 将 y 轴的标签设为 people 的元素值
ax.set_yticklabels(people)
# 设置 y 轴的标签顺序,从上到下
ax.invert_yaxis()
# 设置 x 轴的标签内容
ax.set_xlabel('Performance')
# 设置整个图形的标题
ax.set_title('标题')
# 绘制水平条形图
plt.show()
```

输出如图 5-6 所示。

【实例 5-6】 实现一个堆叠条形图。

本实例通过导入 Matplotlib 库,使用 mpl.rcParams 设置字体样式,使用 subplots 函数返回一个 figure 图像和子图 ax 的 array 列表,使用 bar 函数配置堆叠条形图,使用 set_ylabel 函数设置 y 轴的标签内容,使用 set_title 函数设置图例标题内容,使用 legend 函数显示图例,show 函数用于呈现图形。

```python
# -*- coding: utf-8 -*-
import matplotlib.pyplot as plt
import matplotlib as mpl
# 汉字字体,优先使用楷体,如果找不到楷体,则使用黑体
mpl.rcParams['font.sans-serif'] = ['KaiTi', 'SimHei', 'FangSong']
# 字体大小
mpl.rcParams['font.size'] = 12
# 正常显示负号
mpl.rcParams['axes.unicode_minus'] = False
# 设置标签内容和个数
labels = ['组1', '组2', '组3', '组4', '组5']
# 设置男性分数
men_means = [33, 36, 28, 25, 29]
# 设置女性分数
women_means = [21, 29, 31, 21, 26]
```

```
# 设置男性标准差
men_std = [3, 4, 3, 2, 3]
# 设置女性标准差
women_std = [2, 4, 3, 5, 2]
# 设置条形宽度
width = 0.5
# subplots()函数返回一个figure图像和子图ax的array列表
fig, ax = plt.subplots()
# 使用bar函数配置堆叠条形图。y轴传入labels，x轴传入men_means和women_means，y
轴方向的数据点误差计算方法传入men_std和women_std，布局bottom传入men_means
ax.bar(labels, men_means, width, yerr=men_std, label='男')
ax.bar(labels, women_means, width, yerr=women_std, bottom=men_means,
       label='女')
# 设置y轴的标签内容
ax.set_ylabel('各组分数')
# 设置图形标题内容
ax.set_title('各组和性别得分')
# 显示图例
ax.legend()
# 显示图形
plt.show()
```

输出如图 5-7 所示。

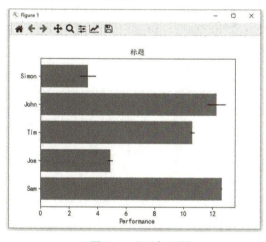

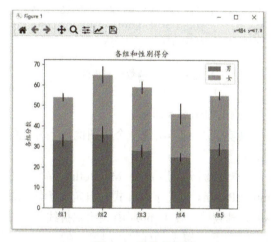

图 5-6　水平条形图　　　　　　　　　图 5-7　堆叠条形图

5.1.4 折线图

折线图呈现随时间变化的连续数据（适合二维的大数据集），通常用于展示在固定时间间隔中数据的变化趋势。

折线图 plot 函数的

```
plot(x,y,label ="标签",color = None,linestyle ="-",linewidth=5,alpha=0.5)
```

参数详解见表 5-3。

表 5-3 折线图 plot 函数参数详解

常用参数名称	功能描述
x, y	x 轴与 y 轴对应的数据,,接收 array
color	（可选）折线的颜色
marker	（可选）折线上数据点的样式
linestyle	（可选）折线的样式，默认"-"
linewidth	（可选）折线的宽度，默认 1
alpha	（可选）折线的透明度
label	（可选）折线的图例
xlabel	（可选）x 轴的标签
ylabel	（可选）y 轴的标签
title	（可选）折线图的标题
legend	（可选）图例的位置
grid	（可选）折线图中绘制网格效果

【实例 5-7】 实现一个基本的折线图。

本实例通过导入 Matplotlib 库，自定义折线图的 x 和 y 轴数据 x_data 和 y_data，使用 plot 函数绘制折线图，show 函数用于呈现图形。

```
import matplotlib.pyplot as plt
# 设置折线图的 x 和 y 轴数据
x_data = ['2011', '2012', '2013', '2014', '2015', '2016', '2017']
y_data = [58000, 60200, 63000, 71000, 84000, 90500, 107000]
# 使用 plot 绘制折线图
plt.plot(x_data, y_data)
# 使用 show 函数呈现折线图
plt.show()
```

输出如图 5-8 所示。

【实例 5-8】 实现一个带图例数字标记的折线图。

本实例通过导入 Matplotlib 库，使用 mpl.rcParams 设置字体样式，自定义折线图的 x 和 y 轴数据，使用 plot 函数绘制折线图，使用 xticks 函数设置 x 轴标签倾斜度，使用 xlabel 函数设置 x 轴标签内容，使用 ylabel 函数设置 y 轴标签内容，使用 title 函数设置图例标题内容，使用 legend 函数显示图例，使用 savefig 函数保存生成的图形，show 函数用于呈现图形。

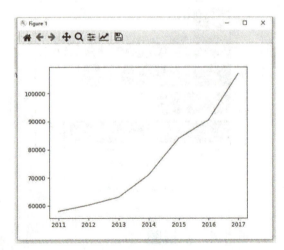

图 5-8 基本折线图

```
# -*- coding: utf-8 -*-
import matplotlib.pyplot as plt
import matplotlib as mpl
# 汉字字体,优先使用楷体,如果找不到楷体,则使用黑体
mpl.rcParams['font.sans-serif'] = ['KaiTi', 'SimHei', 'FangSong']
# 字体大小
mpl.rcParams['font.size'] = 12
```

```
# 正常显示负号
mpl.rcParams['axes.unicode_minus'] = False
# 自定义 x 和 y 轴数据
x = [5, 6, 7, 8]
y = [15, 60, 25, 120]
# "g" 表示绿色,ms 用来设置 o 的大小
plt.plot(x, y, "g", marker='o', ms=30, label="标签")
plt.xticks(rotation=45)
plt.xlabel("日期")
plt.ylabel("数量")
plt.title("标题")
# 使用 upper center 将图例 "标签" 显示到顶部中间
plt.legend(loc="upper center")
# 在折线图上显示具体数值,ha 参数控制水平对齐方式,va 控制垂直对齐方式
for x1, y1 in zip(x, y):
    plt.text(x1, y1 + 1, str(y1), ha='center', va='bottom', fontsize=30, rotation=0)
# 保存生成的折线图 a.jpg
plt.savefig("a.jpg")
# 显示折线图
plt.show()
```

输出如图 5-9 所示。

【实例 5-9】 实现一个双折线图。

本实例通过导入 Matplotlib 库,使用 mpl.rcParams 设置字体样式,自定义折线图的 x 和 y 轴数据,使用 figure 函数设置图形大小,使用 plot 函数设置两条折线图的 x 和 y 轴,标签和颜色以及线条样式,使用 xticks 函数设置 x 轴刻度,使用 grid 设置网格线样式,使用 legend 函数显示图例,show 函数用于呈现图形。

```
# -*- coding: utf-8 -*-
from matplotlib import pyplot as plt
import matplotlib as mpl
# 汉字字体,优先使用楷体,如果找不到楷体,则使用黑体
mpl.rcParams['font.sans-serif'] = ['KaiTi', 'SimHei', 'FangSong']
# 字体大小
mpl.rcParams['font.size'] = 12
# 正常显示负号
mpl.rcParams['axes.unicode_minus'] = False
# 自定义双折线图 y 轴参数
y_1 = [1, 0, 1, 1, 2, 4, 3, 2, 3, 4, 4, 5, 6, 5, 4, 3, 3, 1, 1, 1]
y_2 = [1, 0, 3, 1, 2, 2, 3, 3, 2, 1, 2, 1, 1, 1, 1, 1, 1, 1, 1, 1]
# 自定义双折线图 x 轴参数
x = range(20, 40)
# 设置图形大小
plt.figure(figsize=(10, 8), dpi=80)
# 设置两条折线图的 x 和 y 轴,标签和颜色以及线条样式
plt.plot(x, y_1, label="小明", color="red")
plt.plot(x, y_2, label="小花", color="blue", linestyle="--")

# 设置 x 轴刻度
_xtick_labels = ["{}年".format(i) for i in x]
plt.xticks(x, _xtick_labels)
# 绘制网格,网格线样式为':'
plt.grid(alpha=0.4, linestyle=':')
# 使用 upper center 将图例 "标签" 显示到顶部中间
plt.legend(loc="upper center")
```

```
# 展示
plt.show()
```

输出如图 5-10 所示。

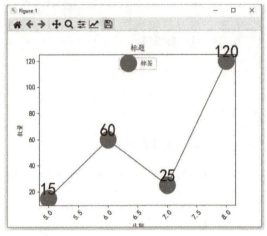
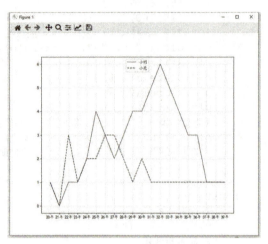

图 5-9　带图例数字标记的折线图　　　　　图 5-10　双折线图

5.1.5　饼图

饼图呈现一系列整体数据中各个部分数据的多少或大小在总体数据的占比状态。

饼图 pie 函数的定义是：

```
pie(x, explode=None, labels=None, colors=None, autopct=None, pctdistance=0.6,
shadow=False, labeldistance=1.1, startangle=None, radius=None, counterclock=True,
wedgeprops=None, textprops=None, center=(0,0), frame=False, hold=None, data=None)
```

参数详解见表 5-4。

表 5-4　饼图 pie 函数参数详解

常用参数名称	功能描述
x	用于绘制饼图的数据，接收 array
explode	(可选)每个扇形离圆心的距离，接收 array
labels	(可选)每个扇形的名称，接收 array
colors	(可选)每个扇形的颜色，接收 array
autopct	(可选)数值的显示方式，接收 string
radius	(可选)饼图的半径，接收 float
shadow	(可选)是否显示阴影，True 或 False

【实例 5-10】实现一个基本的饼图。

本实例通过导入 Matplotlib 库，使用变量 labels 设置饼图各部分数据的标签，使用变量 sizes 设置饼图各部分数据的占比，使用变量 colors 设置饼图各部分数据的颜色，使用变量 explode 设置扇形的间距，使用 pie 函数绘制饼图，使用 axis 函数设置饼状图为正圆，show 函数用于呈现图形。

```
# -*- coding: utf-8 -*-
import matplotlib.pyplot as plt
# 设置饼图各部分数据的标签
labels = 'A', 'B', 'C', 'D'
# 设置饼图各部分数据的占比多少，总体为100
sizes = [15, 30, 45, 10]
# 设置饼图各部分数据的颜色
colors = ['yellowgreen', 'gold', 'lightskyblue', 'lightcoral']
# 展开第二个扇形，与其他扇形的间距为0.1
explode = (0, 0.1, 0, 0)
# 使用 pie 绘制饼图，startangle 控制饼状图的旋转方向，shadow=True 表示是否开启阴影效果
# autopct='%1.1f%%'设置饼图上显示数据，最外面有两个%(转化为百分比，里面 1.1f%即保留一位小数加%
plt.pie(sizes, explode=explode, labels=labels, colors=colors, autopct='%1.1f%%', shadow=True, startangle=90)
# axis('equal')保证饼状图是正圆，否则会有一点角度偏斜
plt.axis('equal')
# 使用 show 函数呈现饼图
plt.show()
```

输出如图 5-11 所示。

【**实例 5-11**】 实现一个环形饼图。

本实例通过导入 Matplotlib 库，使用 mpl.rcParams 设置字体样式，使用 subplots 函数返回一个 figure 图像和子图 ax 的 array 列表，使用 get_cmap 函数设置饼图颜色，使用 cmap 函数设置外层环形饼图颜色，使用 pie 函数设置外层环形饼图参数、半径、颜色、边界宽度和颜色，使用 set 函数设置饼图标题，show 函数用于呈现图形。

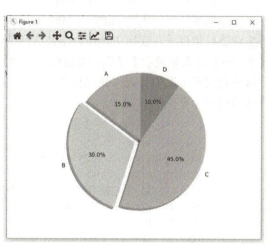

图 5-11 基本饼图

```
# -*- coding: utf-8 -*-
import numpy as np
from matplotlib import pyplot as plt
import matplotlib as mpl
# 汉字字体，优先使用楷体，如果找不到楷体，则使用黑体
mpl.rcParams['font.sans-serif'] = ['KaiTi', 'SimHei', 'FangSong']
# 字体大小
mpl.rcParams['font.size'] = 12
# 正常显示负号
mpl.rcParams['axes.unicode_minus'] = False
# 使用 subplots()函数返回一个 figure 图像和子图 ax 的 array 列表
fig, ax = plt.subplots()
# 设置环形饼图的大小
size = 0.2
# 自定义环形饼图内外两层的显示参数
vals = np.array([[30., 42.], [57., 30.], [19., 20.]])
```

```python
# 使用 get_cmap 设置饼图颜色
cmap = plt.get_cmap("tab20c")
# 设置外层环形饼图颜色
outer_colors = cmap(np.arange(3)*4)
# 设置内层环形饼图颜色
inner_colors = cmap([9, 3, 4, 5, 8,6 ])
# 设置外层环形饼图参数、半径、颜色、边界宽度和颜色
ax.pie(vals.sum(axis=1), radius=1, colors=outer_colors,
       wedgeprops=dict(width=size, edgecolor='w'))
# 设置内层环形饼图参数、半径、颜色、边界宽度和颜色
ax.pie(vals.flatten(), radius=1-size, colors=inner_colors,
       wedgeprops=dict(width=size, edgecolor='w'))
# 设置环形饼图的标题
ax.set(aspect="equal", title='环形饼图')
# 显示环形饼图
plt.show()
```

输出如图 5-12 所示。

【**实例 5-12**】 实现一个极坐标饼图。

本实例通过导入 NumPy 和 Matplotlib 库，使用 mpl.rcParams 设置字体样式，使用 seed 函数生成随机数种子，使用 linspace 函数线性间隔生成 20 个数据，使用 rand 函数分别随机生成 20 个 0~1 的数据表示半径值和宽度值，使用 viridis 函数生成颜色值，使用 subplot 函数的三个参数分别代表子图的行数、列数和第 1 个子图，使用 bar 函数生成极坐标饼图内的条形图，show 函数用于呈现图形。

```python
# -*- coding: utf-8 -*-
import numpy as np
from matplotlib import pyplot as plt
import matplotlib as mpl
# 汉字字体，优先使用楷体，如果找不到楷体，则使用黑体
mpl.rcParams['font.sans-serif'] = ['KaiTi', 'SimHei', 'FangSong']
# 字体大小
mpl.rcParams['font.size'] = 12
# 正常显示负号
mpl.rcParams['axes.unicode_minus'] = False

# 设置随机数种子
np.random.seed(20210801)

# 自定义生成 20 个数
N = 20
# 使用 np.linspace 函数线性间隔生成 20 个数据
theta = np.linspace(0.0, 2 * np.pi, N, endpoint=False)
# 随机生成 20 个 0~1 的数据表示半径值
radii = 5 * np.random.rand(N)
# 随机生成 20 个 0~1 的数据表示宽度值
width = np.pi / 4 * np.random.rand(N)
# 生成颜色值
colors = plt.cm.viridis(radii / 5.)
# subplot 函数 3 个参数分别代表子图的行数、列数和第 1 个子图
ax = plt.subplot(1,1,1, projection='polar')
```

```
# bar 函数表示生成极坐标饼图内的条形图
ax.bar(theta, radii, width=width, bottom=0.0, color=colors, alpha=0.5)
# 显示极坐标饼图
plt.show()
```

输出如图 5-13 所示。

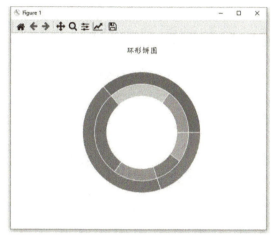

图 5-12　环形饼图

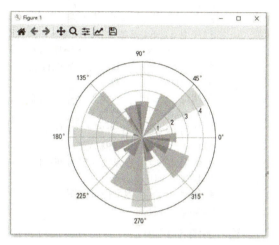

图 5-13　极坐标饼图

5.1.6　直方图

直方图指由一系列高度不等的纵向条纹或线段表示数据分布的情况。一般用横轴表示数据类型，纵轴表示分布情况。

直方图 hist 函数的定义是：

```
hist(array,bins=None,range=None,density=None,weights=None,cumulative=False,
bottom=None,histtype='bar',align='mid',orientation='vertical',rwidth=None,log=False,
color=None,label=None,stacked=False,normed=None,data=None,**kwargs)
```

参数详解见表 5-5。

表 5-5　直方图 plot 函数参数详解

常用参数名称	功能描述
left	x 轴数据，接收 array
height	x 轴所代表数据的数量，接收 array
width	（可选）直方图宽度，接收[0～1]之间的 float
color	（可选）直方图颜色，接收特定 string 或者包含颜色字符串的 array

【实例 5-13】　实现一个基本的直方图。

本实例通过导入 NumPy 和 Matplotlib 库，使用 mpl.rcParams 设置字体样式，使用 randint 函数生成 0～100 之间的 100 个数据，使用 arange 函数设置连续的边界值，使用 hist 函数绘制直方图，使用 xlabel 函数设置 x 轴的标签，使用 ylabel 函数设置 y 轴的标签，使用 title 函数设置直方图标题，使用 xlim 函数设置 x 轴分布范围，show 函数用于呈现图形。

```
# -*- coding: utf-8 -*-
import matplotlib.pyplot as plt
```

```
import numpy as np
plt.rcParams['font.sans-serif']=['SimHei']
#生成0～100之间的100个数据
x=np.random.randint(0,100,100)
#设置连续的边界值，即直方图的分布区间[0,10]，[10,20]
bins=np.arange(0,101,10)
# 使用hist函数绘制直方图
# 使用alpha函数设置透明度，0为完全透明
plt.hist(x,bins,color='g',alpha=0.5)
# 使用xlabel函数设置x轴的标签
plt.xlabel('分数')
# 使用ylabel函数设置y轴的标签
plt.ylabel('计数')
# 使用title函数设置直方图标题
plt.title("直方图标题")
# 使用xlim函数设置x轴分布范围
plt.xlim(0,100)
# 使用show函数呈现直方图
plt.show()
```

输出如图 5-14 所示。

【**实例 5-14**】 实现一个双直方图。

本实例通过导入 NumPy 和 Matplotlib 库，使用 seed 函数生成随机数种子，使用 normal 函数生成正态分布的数据 x1 和 x2，使用 hist 函数设置直方图 n1 和 n2，show 函数用于呈现图形。

```
# -*- coding: utf-8 -*-
import numpy as np
import matplotlib.pyplot as plt
# 设置随机数种子
np.random.seed(20210801)
# 设置均值和标准差值
mean1, sigma1 = 200, 15
mean2, sigma2 = 90, 15
# 使用normal函数生成正态分布的数据。这里传入均值、标准差和个数
x1 = np.random.normal(mean1,sigma1,20000)
x2 = np.random.normal(mean2,sigma2,20000)
# bins：将数据分成80和40组
# color：颜色；alpha：透明度
# density：是密度而不是具体数值
n1, bins1, patches1 = plt.hist(x1, bins=80, density=True, color='black', alpha=1)
n2, bins2, patches2 = plt.hist(x2, bins=40, density=True, color='blue', alpha=0.2)
# 显示双直方图
plt.show()
```

输出如图 5-15 所示。

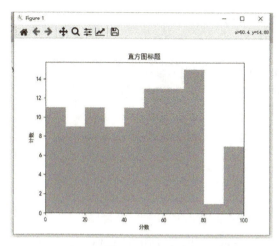

图 5-14　直方图

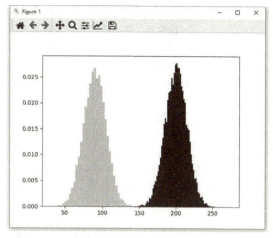

图 5-15　双直方图

【实例 5-15】　实现一个散点直方图。

本实例通过导入 NumPy 和 Matplotlib 库，使用 seed 函数生成随机数种子，使用 randn 函数生成 x 和 y 轴随机数，自定义函数 scatter_hist 用于生成散点直方图，使用 figure 函数设置图形大小，使用 add_axes 函数添加散点图 rect_scatter、水平直方图 rect_histx、垂直直方图 rect_histy 到各子图当中，show 函数用于呈现图形。

```
import numpy as np
import matplotlib.pyplot as plt

# Fixing random state for reproducibility
np.random.seed(20210801)

# 创建 x 和 y 轴随机数
x = np.random.randn(2000)
y = np.random.randn(2000)

# 自定义函数 scatter_hist 用于生成散点直方图
def scatter_hist(x, y, ax, ax_histx, ax_histy):
    # 设置直方图 x 和 y 轴，labelbottom=False 表示底部标签不显示，labelleft=False
表示左部标签不显示
    ax_histx.tick_params(axis="x", labelbottom=False)
    ax_histy.tick_params(axis="y", labelleft=False)

    # 在子图中绘制散点图
    ax.scatter(x, y)

    # 设置直方图数据分组显示的步长
    binwidth = 0.05
    # 设置直方图数据显示范围
    xymax = max(np.max(np.abs(x)), np.max(np.abs(y)))
    lim = (int(xymax/binwidth) + 1) * binwidth
    # 设置数据分组的区间和步长
```

```
    bins = np.arange(-lim, lim + binwidth, binwidth)
    # 设置两个直方图的数据分组个数
    ax_histx.hist(x, bins=bins)
    # orientation='horizontal'表示水平直方图，默认是垂直直方图
    ax_histy.hist(y, bins=bins, orientation='horizontal')

# 设置左边距和宽度
left, width = 0.1, 0.55
# 设置底部边距和高度
bottom, height = 0.1, 0.55
# 设置各子图之间的距离
spacing = 0.01

# 设置散点图 rect_scatter，水平直方图 rect_histx，垂直直方图 rect_histy 的参数
rect_scatter = [left, bottom, width, height]
rect_histx = [left, bottom + height + spacing, width, 0.2]
rect_histy = [left + width + spacing, bottom, 0.2, height]

# 设置宽 8、长 8（单位为 inch）布局
f ig = plt.figure(figsize=(8, 8))
# 添加散点图 rect_scatter，水平直方图 rect_histx，垂直直方图 rect_histy 到各子图当中
ax = fig.add_axes(rect_scatter)
ax_histx = fig.add_axes(rect_histx, sharex=ax)
ax_histy = fig.add_axes(rect_histy, sharey=ax)

# 使用自定义函数 scatter_hist，并传入参数
scatter_hist(x, y, ax, ax_histx, ax_histy)
# 显示散点直方图
plt.show()
```

输出如图 5-16 所示。

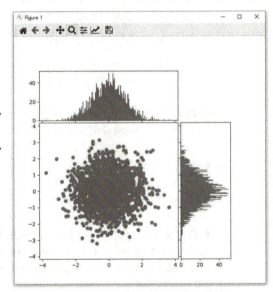

图 5-16　散点直方图

5.1.7 箱形图

箱形图（又称箱线图、盒须图）是通过显示一组数据的最大值、最小值、中位数和上下四分位数显示数据分布情况的统计图。

箱形图 boxplot 函数的定义是：

```
boxplot(x, notch=None, labels=None,meanline=None, sym=None, vert=None, whis=None, positions=None, widths=None,** kwargs)
```

参数详解见表 5-6。

表 5-6　箱形图 boxplot 函数参数详解

常用参数名称	功能描述
x	绘制箱形图的数据，接收 array
notch	（可选）中间箱体是否有缺口，接收 boolean
labels	（可选）每一个箱形图的标签，接收 array
meanline	（可选）是否显示均值线，接收 boolean
widths	（可选）每个箱体的宽度，接收 array
vert	（可选）图形是纵向或者横向，接收 boolean

【实例 5-16】 实现一个基本的箱形图。

本实例通过导入 NumPy、Pandas 和 Matplotlib 库，使用 seed 函数生成随机数种子，使用 randn 函数生成 0～1 之间的 5 行 4 列的数据，使用 boxplot 函数绘制箱形图，show 函数用于呈现图形。

```
import numpy as np
import matplotlib.pyplot as plt
import pandas as pd
#设置随机种子
np.random.seed(2)
# 使用 np.random.rand 生成 0～1 之间的
5 行 4 列的数据
x = np.random.rand(5,4)
# 使用 boxplot 函数绘制箱形图
plt.boxplot(x,widths=0.3,labels=
['A', 'B', 'C', 'D'])
# 使用 show 函数呈现箱形图
plt.show()
```

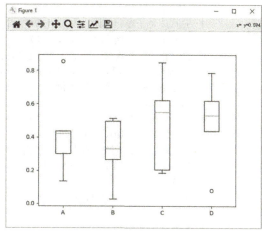

图 5-17　箱形图

输出如图 5-17 所示。

5.2　组合图

Matplotlib 可以将多种图形自定义地布局到一张画布上面，实现多图组合的呈现方式。

5.2.1　曲线组合图

【实例 5-17】 实现一个曲线组合图。

本实例通过导入 NumPy 和 Matplotlib 库，使用 rcParams 设置字体格式，linspace 函数设置 0～2*np.pi 之间 200 个数据样本，sin 函数作为 y1 的数据，cos 函数作为 y2 的数据，sin 函数作为 y3 的数据，subplot 函数设置子图布局，plot 函数绘制组合图，legend 函数呈现图例，show 函数用于呈现图形。

```
# -*- coding: utf-8 -*-
import numpy as np
import matplotlib.pyplot as plt
import matplotlib as mpl
```

```
# 汉字字体，优先使用楷体，如果找不到楷体，则使用黑体
mpl.rcParams['font.sans-serif'] = ['KaiTi', 'SimHei', 'FangSong']
# 字体大小
mpl.rcParams['font.size'] = 12
# 正常显示负号
mpl.rcParams['axes.unicode_minus'] = False
# 创建 0～2*np.pi 之间 200 个数据样本
x = np.linspace(0, 2*np.pi, 200)
# 设置字体大小为 12
plt.rc('font', size=12)
# 使用 sin 正弦函数作为 y1 的数据
y1 = np.sin(x)
# 使用 cos 余弦函数作为 y2 的数据
y2 = np.cos(x)
# 使用 sin 正弦函数作为 y3 的数据
y3 = np.sin(x**2)
# 左上 1 号子窗口图形
ax1 = plt.subplot(2, 2, 1)
# 使用 plot 函数绘制左上 1 号子窗口图形
ax1.plot(x, y1, 'r', label='$sin(x)$')
# 使用 legend 函数呈现图例
plt.legend()
# 右上 2 号子窗口
ax2 = plt.subplot(2, 2, 2)
# 使用 plot 函数绘制右上 2 号子窗口图形
ax2.plot(x, y2, 'b--', label='$cos(x)$')
# 使用 legend 函数呈现图例
plt.legend()
# 底部 1 号子窗口
ax3 = plt.subplot(2, 1, 2)
# 使用 plot 绘制下面 1 号子窗口图形
ax3.plot(x, y3, 'k--', label='$sin(x^2)$')
# 使用 legend 函数呈现图例
plt.legend()
# 使用 show 函数呈现组合图
plt.show()
```

输出如图 5-18 所示。

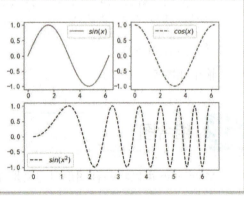

图 5-18　曲线组合图

5.2.2　柱状、散点、折线组合图

【**实例 5-18**】 实现一个柱状、散点、折线组合图。

本实例通过导入 NumPy 和 Matplotlib 库，使用 rcParams 设置字体格式，变量 data 设置组合图显示的数据名字和大小，subplots 函数绘制 1 行 3 列，宽 9，长 3 的布局，bar 函数配置柱状图，scatter 函数配置散点图，plot 函数配置折线图，suptitle 函数设置子标题，show 函数用于呈现图形。

```
# -*- coding: utf-8 -*-
import matplotlib.pyplot as plt
```

```python
import matplotlib as mpl
# 汉字字体，优先使用楷体，如果找不到楷体，则使用黑体
mpl.rcParams['font.sans-serif'] = ['KaiTi', 'SimHei', 'FangSong']
# 字体大小
mpl.rcParams['font.size'] = 12
# 正常显示负号
mpl.rcParams['axes.unicode_minus'] = False
# 设置组合图显示的数据名字和大小。
data = {'鼠标': 20, '键盘': 35, '显示器': 15, '机箱': 10}
names = list(data.keys())
values = list(data.values())
# 使用 subplots 函数绘制 1 行 3 列, 宽 9, 长 3 的布局, sharey=True 表示 data 的数据都在 y 轴上显示
fig, axs = plt.subplots(1, 3, figsize=(9, 3), sharey=True)
# 给各子图绘制 bar 柱状图、scatter 散点图、plot 折线图
axs[0].bar(names, values)
axs[1].scatter(names, values)
axs[2].plot(names, values)
# 设置子标题
fig.suptitle('柱状、散点、折线组合图')
# 显示组合图
plt.show()
```

输出如图 5-19 所示。

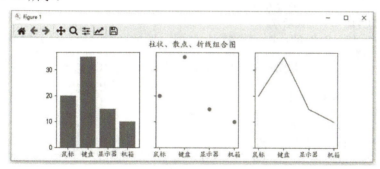

图 5-19　柱状、散点、折线组合图

5.2.3　直方图组合图

【实例 5-19】　实现一个直方图组合图。

本实例通过导入 NumPy 和 Matplotlib 库，使用 rcParams 设置字体格式，seed 函数生成随机数种子，randn 函数生成 1000 个随机数，步长为 3，subplots 函数设置 2 行 2 列的布局，分别装载 4 子图 ax0、ax1、ax2 和 ax3，hist 函数设置直方图的显示方式，legend 函数设置图例显示大小，set_title 函数设置标题，tight_layout 函数实现自动调整子图参数，使之填充整个图像区域，show 函数用于呈现图形。

```python
# -*- coding: utf-8 -*-
import numpy as np
import matplotlib.pyplot as plt
import matplotlib as mpl
# 汉字字体，优先使用楷体，如果找不到楷体，则使用黑体
```

```python
mpl.rcParams['font.sans-serif'] = ['KaiTi', 'SimHei', 'FangSong']
# 字体大小
mpl.rcParams['font.size'] = 12
# 正常显示负号
mpl.rcParams['axes.unicode_minus'] = False
# 设置随机数种子
np.random.seed(19680801)
# 设置分组数
n_bins = 10
# 生成 1000 个随机数,步长为 3
x = np.random.randn(1000, 3)
# 使用 subplots 设置 2 行 2 列的布局,分别装载 4 子图 ax0、ax1、ax2 和 ax3,其中第一行放入 ax0, ax1,第二行放入 ax2, ax3
fig, ((ax0, ax1), (ax2, ax3)) = plt.subplots(nrows=2, ncols=2)
# 设置显示颜色
colors = ['red', 'tan', 'lime']
# 设置直方图 ax0 的显示方式
ax0.hist(x, n_bins, density=True, histtype='bar', color=colors, label=colors)
# 设置图例显示大小
ax0.legend(prop={'size': 8})
# 设置标题
ax0.set_title('带图例的直方图 ax0')
# 设置直方图 ax1 的显示方式
ax1.hist(x, n_bins, density=True, histtype='bar', stacked=True)
# 设置标题
ax1.set_title('堆叠直方图 ax1')
# 设置直方图 ax2 的显示方式
ax2.hist(x, n_bins, histtype='step', stacked=True, fill=False)
# 设置标题
ax2.set_title('未填充的堆叠直方图')

# 自定义直方图 ax3 的数据
x_multi = [np.random.randn(n) for n in [12000, 5500, 2300]]
# 设置直方图 ax3 的显示方式
ax3.hist(x_multi, n_bins, histtype='bar')
# 设置标题
ax3.set_title('多数据直方图 ax3')
# 自动调整子图参数,使之填充整个图像区域
fig.tight_layout()
# 显示组合图
plt.show()
```

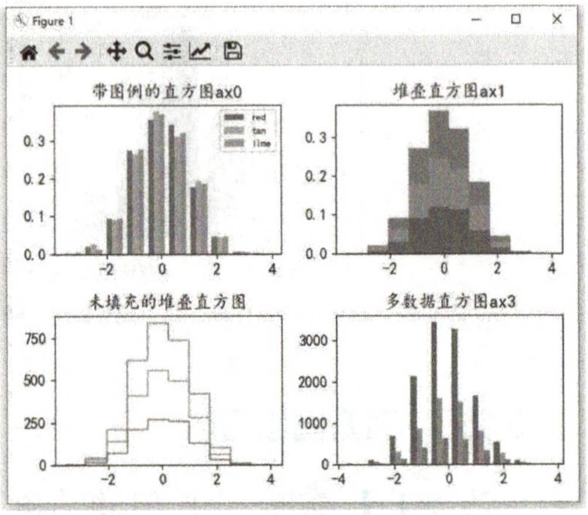

输出如图 5-20 所示。

图 5-20 直方图组合图

【任务实施】

任务 1 使用饼图实现零售总额数据分析

分析重庆 2020 年 12 月线上单位商品零售的数据(见图 5-21)特点,使用 Matplotlib 饼图实现零售总额数据分析。

1)导入指定的库。

```python
# 导入 Pandas 库,并将其命名为 pd
```

```
import pandas as pd
# 导入Matplotlib可视化库
import matplotlib.pyplot as plt
# 设置指定的字体，防止乱码
plt.rcParams['font.sans-serif']=['SimHei']
plt.rcParams['axes.unicode_minus']=False
```

2）获取指定字段的数据：两个字段“指标”和“2020 年 12 月”，如图 5-22 所示。

```
data = pd.read_csv("bingtu.csv")
data
```

	A	B
1	指标	2020年12月
2	粮油、食品类	620.11
3	饮料类	53.04
4	烟酒类	82.19
5	服装鞋帽、针、纺织品类	249.29
6	化妆品类	40
7	金银珠宝类	47.27
8	日用品类	200.87

图 5-21　重庆 2020 年 12 月线上单位商品零售的数据

	指标	2020年12月
0	粮油、食品类	620.11
1	饮料类	53.04
2	烟酒类	82.19
3	服装鞋帽、针、纺织品类	249.29
4	化妆品类	40.00
5	金银珠宝类	47.27
6	日用品类	200.87

图 5-22　获取“指标”和“2020 年 12 月”字段的数据

3）使用变量 data_index 接收获取字段为“指标”的值，数据类型为 object，如图 5-23 所示。

```
data_index = data["指标"]
data_index
```

4）使用变量 data_dec 接收获取字段为“2020 年 12 月”的值，数据类型为 float64，如图 5-24 所示。

```
data_dec = data["2020年12月"]
data_dec
```

```
0          粮油、食品类
1             饮料类
2             烟酒类
3     服装鞋帽、针、纺织品类
4            化妆品类
5           金银珠宝类
6            日用品类
Name: 指标, dtype: object
```

图 5-23　获取“指标”字段的详细信息

```
0    620.11
1     53.04
2     82.19
3    249.29
4     40.00
5     47.27
6    200.87
Name: 2020年12月, dtype: float64
```

图 5-24　获取“2020 年 12 月”字段的详细信息

5）图形配置和呈现。

```
# 设置画布的尺寸
plt.figure(figsize=(6,6))
# 设定各项距离圆心的半径
explode=[0.01,0.01,0.01,0.01,0.01,0.01,0.01]
# 绘制饼图，并传入参数
plt.pie(data_dec,explode=explode,labels=data_index,autopct='%1.1f%%')
# 绘制标题
plt.title('2020年12月重庆各类指标饼图')
# 呈现饼图
```

```
plt.show()
```

输出如图 5-25 所示。

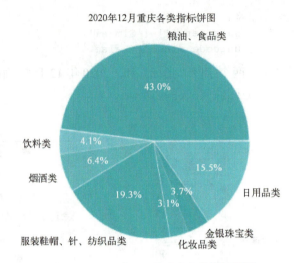

图 5-25　2020 年 12 月重庆各类指标饼图

饼图可视化分析：2020 年 12 月重庆市实现零售总额 1290.77 亿元。粮油、食品类收入占比最大 620.11 亿元，占比总消费 48%，说明重庆人民对于美食的需求依然强劲。

消费升级类商品占比较大，服装鞋帽售额占总消费比次之，为 249.29 亿元，占比总消费 19.3%，接下来是日用品类，为 200.87 亿元，占比总消费 15.5%。化妆品类和金银珠宝类旗鼓相当，分别为 40 亿元和 47.27 亿元。

任务 2　使用折线图实现零售总额数据分析

分析重庆 2020 年 2～12 月日用品累计增长的数据特点，使用 Matplotlib 折线图实现日用品累计增长的数据分析，如图 5-26 所示：

1）导入指定的库。

```
# 导入 Pandas 库，并将其命名为 pd
import pandas as pd
# 导入 Matplotlib 可视化库
import matplotlib.pyplot as plt
# 设置指定的字体，防止乱码
plt.rcParams['font.sans-serif']=['SimHei']
plt.rcParams['axes.unicode_minus']=False
```

2）导入数据，并观察分析两个字段"日期"和"日用品累计增长(%)"，如图 5-27 所示。

```
data = pd.read_csv("demo1.csv")
data
```

3）使用变量 data_date 接收获取字段为"日期"的值，数据类型为 object，并倒序输出，如图 5-28 所示。

```
data_date = data["日期"][::-1]
data_date
```

项目 5　Matplotlib 数据分析可视化库

	A	B
	日期	日用品累计增长(%)
	2020年12月	19.6
	2020年11月	16.5
	2020年10月	12.1
	2020年9月	5
	2020年8月	3.3
	2020年7月	0.1
	2020年6月	-2
	2020年5月	-8.5
	2020年4月	-12.7
	2020年3月	-13.6
	2020年2月	-28.7

图 5-26　重庆 2020 年 2 月到 12 月日用品累计增长

	日期	日用品累计增长(%)
0	2020年12月	19.6
1	2020年11月	16.5
2	2020年10月	12.1
3	2020年9月	5.0
4	2020年8月	3.3
5	2020年7月	0.1
6	2020年6月	-2.0
7	2020年5月	-8.5
8	2020年4月	-12.7
9	2020年3月	-13.6
10	2020年2月	-28.7

图 5-27　获取"日期"和"日用品累计增长(%)"字段的数据

4）使用变量 data_detail 接收获取字段为"日用品累计增长(%)"的值，数据类型为 float64，并倒序输出，如图 5-29 所示。

```
data_detail = data["日用品累计增长(%)"][::-1]
data_detail
```

```
10    2020年2月
9     2020年3月
8     2020年4月
7     2020年5月
6     2020年6月
5     2020年7月
4     2020年8月
3     2020年9月
2     2020年10月
1     2020年11月
0     2020年12月
Name: 日期, dtype: object
```

```
10    -28.7
9     -13.6
8     -12.7
7      -8.5
6      -2.0
5       0.1
4       3.3
3       5.0
2      12.1
1      16.5
0      19.6
Name: 日用品累计增长(%), dtype: float64
```

图 5-28　获取"日期"字段的详细信息　　图 5-29　获取"日用品累计增长(%)"字段的详细信息

5）图形配置和呈现。

```
# 使用 plt 的 plot 方法接收数据
plt.plot(data_date,data_detail)
# 设置横坐标 45°呈现
plt.xticks(rotation=45)
# 使用 plt 的 show 方法呈现数据
plt.show()
```

如图 5-30 所示。

折线图可视化分析：从图中可以看出，重庆 2020 年 2 月到 2020 年 12 月日用品累计增长数据是稳步上涨的。

任务 3　使用双柱状图实现零售总额变化情况数据分析

分析重庆 2020 年 2~12 月日用品类和家电类的数据特点，使用 Matplotlib 折线图实现日用品累计增长的数据分析，如图 5-31 所示。

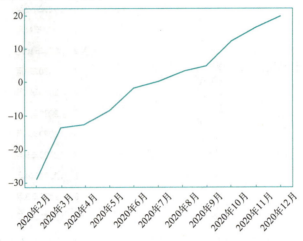

图 5-30　折线图实现日用品累计增长

1）导入指定的库。

```
# 导入 Pandas 库，并将其命名为 pd
import pandas as pd
# 导入 NumPy 库，并将其命名为 np
import numpy as np
# 导入 Matplotlib 可视化库
import matplotlib.pyplot as plt
# 设置指定的字体，防止乱码
plt.rcParams['font.sans-serif']=['SimHei']
plt.rcParams['axes.unicode_minus']=False
```

2）导入数据，并观察分析三个字段"日期""日用品类"和"家用电器类"，如图 5-32 所示。

```
data = pd.read_csv("shuangzhuzhuangtu.csv")
data
```

	A	B	C
	日期	日用品类	家用电器类
	2020年12月	200.87	268.66
	2020年11月	182.71	241.89
	2020年10月	166.66	213.55
	2020年9月	149.72	190.14
	2020年8月	132.78	165.86
	2020年7月	116.86	142.22
	2020年6月	100.73	119.12
	2020年5月	82.06	87.56
	2020年4月	65.75	62.53
	2020年3月	49.5	43.7
	2020年2月	32.95	21.42

图 5-31　重庆 2020 年 2～12 月日用品类和家电类的数据

	日期	日用品类	家用电器类
0	2020年12月	200.87	268.66
1	2020年11月	182.71	241.89
2	2020年10月	166.66	213.55
3	2020年9月	149.72	190.14
4	2020年8月	132.78	165.86
5	2020年7月	116.86	142.22
6	2020年6月	100.73	119.12
7	2020年5月	82.06	87.56
8	2020年4月	65.75	62.53
9	2020年3月	49.50	43.70
10	2020年2月	32.95	21.42

图 5-32　获取"日期""日用品类"和"家用电器类"字段的数据

3）使用变量 data_pro 获取字段"日用品类"的值，并倒序输出，如图 5-33 所示。

```
data_pro = data["日用品类"][::-1]
```

```
data_pro
```

4）使用变量 data_appliance 获取字段"家用电器类"的值，并倒序输出，如图 5-34 所示。

```
data_appliance = data["家用电器类"][::-1]
data_appliance
```

10	32.95		10	21.42
9	49.50		9	43.70
8	65.75		8	62.53
7	82.06		7	87.56
6	100.73		6	119.12
5	116.86		5	142.22
4	132.78		4	165.86
3	149.72		3	190.14
2	166.66		2	213.55
1	182.71		1	241.89
0	200.87		0	268.66

Name: 日用品类, dtype: float64　　Name: 家用电器类, dtype: float64

图 5-33　获取"日用品类"字段的详细信息　　图 5-34　获取"家用电器类"字段的详细信息

5）使用变量 data_date 获取字段"日期"的值，并倒序输出，如图 5-35 所示。

```
data_date = data["日期"][::-1]
data_date
```

10	2020年2月
9	2020年3月
8	2020年4月
7	2020年5月
6	2020年6月
5	2020年7月
4	2020年8月
3	2020年9月
2	2020年10月
1	2020年11月
0	2020年12月

Name: 日期, dtype: object

图 5-35　获取"日期"字段的详细信息

6）图形配置和呈现。

```
labels = data_date
# 使用 np 的 range 函数根据 data_date 的数量设置 x 轴的位置
x = np.arange(len(labels))
# 定义柱状图条的宽度
width = 0.35
# 初始化 plt 为对象 fig 和 ax
fig, ax = plt.subplots()
# 使用 ax 的 bar 方法配置双柱状图
rects1 = ax.bar(x - width/2, data_pro, width, label='日用品类')
rects2 = ax.bar(x + width/2, data_appliance, width, label='家用电器类')

# 设置 y 轴子标签
ax.set_ylabel('零售额（亿元）')
# 设置双柱状图主标签
ax.set_title('2020 年 2～12 月日用品类和家用电器类零售总额变化情况')
# 设置 x 轴的标签位置
```

```
ax.set_xticks(x)
# 设置 x 轴的标签内容和呈现角度
ax.set_xticklabels(labels,rotation=-30)
# 呈现图例
ax.legend()
# 全局设计整体布局
fig.tight_layout()
# 呈现整个双柱状图
plt.show()
```

对双柱状图进行分析可知：2020 年 2~12 月重庆全市实现日用品类和家用电器类零售总额 469.53 亿元，重庆 2020 年 2~12 月日用品和家用电器类数据是稳步上涨的，如图 5-36 所示。

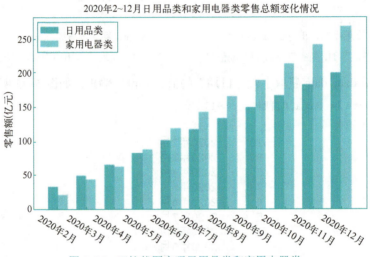

图 5-36　双柱状图实现日用品类和家用电器类

练习题

（1）获取 csv 源数据。
（2）处理和分析数据，实现游戏占比饼图。
（3）根据每年游戏销售数据实现散点图、折线图和柱状图。

项目 6　基于 Hadoop 的数据分析

【项目分析】

本项目旨在理解和掌握 Hadoop 数据分析工具的基本概念和基础语法。具体内容如下。

1）Hadoop 的基本概念、作用、特点和主流发布版本。

2）Hadoop 核心组件的工作原理，包括 HDFS、MapReduce 和 YARN。

3）Hadoop 常用生态组件：Hive、Spark、HBase、Kafka、Flume、Sqoop 和 Zookeeper 的基本概念、组织架构、安装、部署和应用。

4）Hadoop 数据分析案例介绍。

【知识准备】

6.1　掌握 Hadoop 框架和生态组件

6.1.1　Hadoop 简介

Hadoop 是主流的大数据存储和分析平台之一。它是来源于 Apache 基金会以 Java 编写的开源分布式框架项目。其核心组件是 HDFS、YARN 和 MapReduce，其它组件为：HBase、Hive、Zookeeper、Spark、Kafka、Flume、Ambari、和 Sqoop 等。这些组件共同提供了一套完整的服务。Hadoop 可以将大规模海量数据进行分布式并行处理。Hadoop 具有高度容错性、可扩展性、高可靠性和稳定性，如图 6-1 所示。

图 6-1　Hadoop 及其生态圈

目前市面上的 Hadoop 发行版本主要有三个，分别来自 Apache、Cloudera 和 Hortonworks。其中，Apache Hadoop 是 Hadoop 原生版本，比较基础，适合初学者。Cloudera 主要应用于大型的互联网企业。Hortonworks 是完全开源的，在文档方面比较有优势。

6.1.2 Hadoop 核心组件和工作原理

Hadoop 有三个核心组件：HDFS（数据存储）、MapReduce（分布式离线计算）和 YARN（资源调度）。

（1）Hadoop 分布式文件系统（Hadoop Distributed File System，HDFS）

HDFS 属于 Hadoop 的底层核心组件。它是分布式文件系统的一种，并具备以下特点。

1）海量数据和流式数据访问读写交互能力。

由于能够高效地实现数据输出和输入，因此它非常适合建立在大规模数据群之中。它能够以批处理的方式实现对大规模数据的处理，特别是以流式地读写数据进行批量处理。它还能在多平台之间实现海量数据的交互处理。其基本设计思想是：一次写入，多次读取。

2）高度容错能力。

Hadoop 是建立在大规模集群之上，而这些集群中难免有某个单机或组群出现硬件故障。HDFS 充分考虑了此问题，当某个单机或组群出现硬件故障时，HDFS 将通过 Master/Slave 体系结构中的 Namenode 和 Datanode 模式对数据进行控制、管理、存储、创建、删除、映射和复制，实现 Hadoop 大数据平台的自我检测、自我诊断和自我恢复。因此，HDFS 不仅能够解决硬件失败的问题，还能够在异构架构之间复制模块，优化和弥补异构机群之间产生的各种问题，保证整个大数据平台系统的服务质量。

3）移动计算。

HDFS 能够将 Hadoop 大数据平台中各种应用程序以网络为基础，迁移至需要执行计算处理的数据保存地或更近的位置，从而进行高效率、低功耗、移动式的数据处理。

4）部署方便。

由于 HDFS 是用 Java 编写的，因此它很容易部署在大规模廉价的分布式商用集群之上。

HDFS 的工作过程如图 6-2 所示。

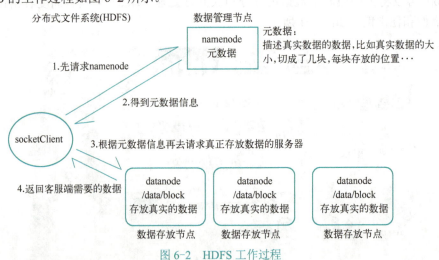

图 6-2　HDFS 工作过程

HDFS 读写数据的工作过程如下。

1）客户端（Client）首先请求 Namenode 中保存的元数据。元数据是描述真实数据的大小和位置等状态信息。客户端读写数据，都需要向 Namenode 发送请求。

2）如果是读取数据，则通过 Namenode 中元数据得到分布式集群中目标数据的真实存放位置后，客户端可以访问指定的集群节点。如果是写数据则通过 Namenode 申请在集群节点上开辟新的存储空间。

3）客户端向指定的集群节点发送请求获取或写入目标数据。

4）如果是读数据则集群节点根据客户端的请求内容发送数据给客户端。如果是写数据则集群节点向 Namenode 报告写入数据的状态，并由 Namenode 统一管理其元数据。

在整个读写数据的过程中，由于所有目标数据的元数据都在 Namenode 中进行统一管理，因此，HDFS 的 Namenode 会对每一个目标数据均衡地安排客户端来读写数据，包括当部分目标数据出现节点失败时，从而能够最优化地实现集群的并发和并行操作。

（2）第一代 MapReduce

MapReduce 是建立在 HDFS 之上的数据映射和化简并行处理技术。它是一种具有线性特质、可扩展的编程模型。它对网络服务器日志等半结构化和非结构化数据的处理非常有效。Map 和 Reduce 分别代表两种函数。前者主要负责将一个任务进行碎片化处理，后者主要负责将各种碎片化信息进行重组汇总。如图 6-3 所示。

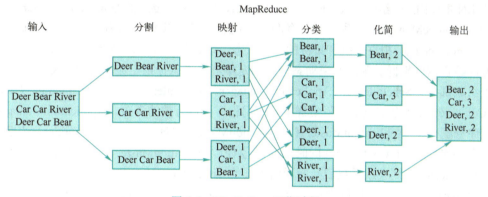

图 6-3　MapReduce 工作过程

MapReduce 的工作过程如下。

1）从数据源获取需要输入的数据。

2）通过 split 函数对原始数据进行分割。

3）通过 map 函数对分割的数据进行标记映射。

4）通过 shuffle 函数对标记映射之后的数据进行分类。

5）通过 reduce 函数对分类之后的数据进行聚类统计。

6）输出 MapReduce 的统计分析结果。

与 HDFS 的 Namenode 和 Datanode 模式相比，第一代 MapReduce 同样采用了 Master/Slave 体系结构，即 Jobtracker 和 Tasktracker 模式。在人机交互过程中，用户提交的工作任务都形成一个 Jobtracker 和许多个等长的 Tasktracker。Jobtracker 负责在工作和任务之间进行合理安排协调。Tasktracker 负责具体执行相应任务安排，并在执行过程中及时向 Jobtracker 返回工作情况（也称心跳）。在数据的并行处理过程中，当某个节点或群组没有心跳反馈时，Jobtracker 即认为该区域不能够提供数据服务，并重新进行工作任务安排。一个 Jobtracker 所划分的 Tasktracker

越多，其在整个大数据平台环境中的并行处理效率和设备资源的合理利用率也就越高。同时，MapReduce 还能够将本地数据进行本地化处理，提高数据访问的效率，如图 6-4 所示。

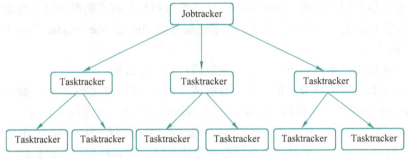

图 6-4　第一代 MapReduce 的 Jobtracker 和 Tasktracker 模式

（3）YARN（Yet Another Resource Negotiator）或第二代 MapReduce

由于第一代 MapReduce 存在一定的局限，例如，Jobtracker 既要负责资源管理，又要监控、跟踪、记录和控制任务，成为整个 MapReduce 的性能瓶颈。最重要的是，第一代 MapReduce 在系统的整体资源利用率方面相对较低。因此，为了优化和提升 MapReduce 的性能和资源利用率，Hadoop 引入了 YARN 专门用于整合 Hadoop 集群资源，并支持其他分布式计算模式。

YARN 主要由三个组件组成：ResourceManager、NodeManager 和 ApplicationMaster。

1）ResourceManager 有系统所有资源的分配权，负责集群中所有应用程序资源的分配。ResourceManager 能够根据系统和集群的真实状态，根据各个应用程序进程优先级和资源利用率最优方面综合调度集群资源，实现动态分配指定集群节点运行指定应用程序。ResourceManager 有两个组成部分：Scheduler 和 ApplicationManager。Scheduler 负责调度 Container（资源容器库），Container 在各个节点上保障集群中各应用程序获得有效的资源。ApplicationManager 为各应用程序分配一个 ApplicationMaster（ApplicationMaster 本身也是一个 Container）来负责通信并管理其他节点上的 Container。同时，负责监控 ApplicationMaster，并在其某个应用程序的 ApplicationMaster 遇到失败时重启其所通信的其他 Container。

2）NodeManager 负责集群中每一个单一节点与 ResourceManager 的工作通信，负责启动和管理单一节点中应用程序 Container 的生命周期，以及监控和跟踪应用程序的资源使用状态。NodeManager 将单个节点的资源情况、工作日志和运行状态数据汇总到 ResourceManager。每一个 NodeManager 都会在 ResourceManager 注册，并通过定期的"心跳"方式发送通信数据给 ResourceManager，保障了 ResourceManager 能够及时掌握整个集群中所有节点的运行状态和资源情况。

3）ApplicationMaster。集群中每一个应用程序都有自己的 ApplicationMaster。ApplicationMaster 负责与 ResourceManager 通信获得指定应用程序运行所需的资源 Container。ApplicationMaster 负责与 Scheduler 通信获得正确的 Container 来运行指定的应用程序。ApplicationMaster 本身作为一个进程和 NodeManager 一起负责在节点和集群中协调、管理和监控各个应用程序。ApplicationMaster 也是通过"心跳"的方式与 ResourceManager 通信，保障了每个 ApplicationMaster 所管理的应用程序能够在 ResourceManager 获得有效的系统资源。同时，也能够让 ResourceManager 及时知道何时可以释放指定节点的资源。

YARN 的工作过程如图 6-5 所示。

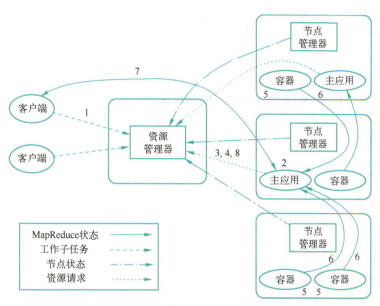

图 6-5　YARN 工作过程

YARN 的工作过程如下。

1）客户端向 ResourceManager 提交使用应用程序的请求。ResourceManager 在接到该请求后会生成一个 ApplicationMaster 对象指向该应用程序并分配一个 application ID 用于标识该应用程序所有的资源信息。

2）ResourceManager 根据集群各节点状态寻找一个能够满足运行当前应用程序所需资源的 NodeManager，并通过其 Container 启动 ApplicationMaster 对象。

3）在 ApplicationMaster 对象启动成功之后，ApplicationMaster 向 ResourceManager 提交注册信息。这样，客户端就可以通过从 ResourceManager 注册表里获得的 ApplicationMaster 状态信息与 ApplicationMaster 直接进行通信了。

4）每个应用程序的 ApplicationMaster 会通过与 ResourceManager 的交互告知 ResourceManager 当前应用程序的资源使用情况，并由 ResourceManager 根据当前系统资源状态，使用相应的资源调度算法为该应用程序的 ApplicationMaster 分配最合适的 Container 资源。

5）ApplicationMaster 被分配到最合适的 Container 资源后，NodeManager 会启动对 Container 的实例化过程。这时，ResourceManager 将脱离该应用程序的进一步执行，只负责与其他 ApplicationMaster 在资源调度和竞争方面的控制。

6）应用程序具体运行的代码会在 Container 中执行。Container 将运行状态信息发送给应用程序对应的 ApplicationMaster。ApplicationMaster 也会通过"心跳"方式将应用程序对于资源请求和释放的信息不断发送给 ResourceManager。

7）前面的 ApplicationMaster 注册之后，客户端可以直接和 ApplicationMaster 进行关于应用程序运行状态的通信。

8）应用程序的进程结束之后，ApplicationMaster 会向 ResourceManager 申请取消注册信息。Resourcemanager 可以命令 NodeManager 杀死或收回 container 的相关资源给系统和清理 container 相关的日志信息。

6.1.3 Hadoop 安装、部署和应用

（1）所需平台和软件
- 三台虚拟机（VMware Workstation）。
- Linux 版本：CentOS Linux release 7.7.1908 (Core)。
- jdk-8u144-linux-x64.tar。
- Hadoop-2.7.2.tar.gz。
- MobaXterm。

（2）虚拟机环境准备

1）修改主机名。

```
#查看主机名
hostname
#永久修改主机名：hostnamectl set-hostname "hostname"
hostnamectl set-hostname Hadoop102
```

2）配置网络。

```
#修改静态 IP 地址
vim /etc/sysconfig/network-scripts/ifcfg-ens33
##如果这里提示 vim 命令不存在，就在线安装：yum install vim
#进入之后修改的内容：

#修改为静态 ip
BOOTPROTO=static
#添加静态 ip 地址
IPADDR=192.168.19.102
#添加网关
GATEWAY=192.168.19.2
#添加 DNS
DNS1=192.168.19.2
DNS2=8.8.8.8

#重启网络
systemctl restart network
```

3）配置 hosts 文件。

```
#修改/etc/hosts 文件
[root@Hadoop102 ~]# vim /etc/hosts

#追加以下内容：
#IP      hostname
192.168.19.100 Hadoop100
192.168.19.101 Hadoop101
192.168.19.102 Hadoop102
192.168.19.103 Hadoop103
192.168.19.104 Hadoop104
```

4）关闭防火墙。

```
#停止 NetworkManager 服务
systemctl stop NetworkManager
```

5）创建 Hadoop 用户。

```
#创建 Hadoop 用户来操作 Hadoop 集群，root 权限太大，应避免使用 root 权限来操作 Hadoop
#添加用户
useradd Hadoop
#更新密码
passwd Hadoop
```

6）赋予 Hadoop 用户管理者权限。

```
#修改 /etc/sudoers 文件，找到 root 一行，在 root 下面添加一行，如下所示
vim /etc/sudoers

## Allow root to run any commands anywhere
root    ALL=(ALL)       ALL
Hadoop  ALL=(ALL)    NOPASSWD:ALL
```

7）使用 Hadoop 用户操作。

```
#切换 Hadoop 用户的方式
[root@Hadoop102 ~]# su - Hadoop
[Hadoop@Hadoop102 ~]$
```

8）在/opt 目录下创建文件夹。

```
#进入 opt 目录
[Hadoop@Hadoop102 ~]$ cd /opt/

#创建 software、module 两个文件夹
[Hadoop@Hadoop102 opt]$ sudo mkdir software module

#修改 software、module 文件夹的所有者为 Hadoop
[Hadoop@Hadoop102 opt]$ sudo chown Hadoop:Hadoop module/ software/
```

（3）卸载旧的 JDK

```
#查询 Linux 的原有 Java
[Hadoop@Hadoop102 opt]$ rpm -qa | grep java
tzdata-java-2019b-1.el7.noarch
python-javapackages-3.4.1-11.el7.noarch

#卸载原有 Java
[Hadoop@Hadoop102 opt]$ sudo rpm -e tzdata-java-2019b-1.el7.noarch
[Hadoop@Hadoop102 opt]$ sudo rpm -e python-javapackages-3.4.1-11.el7.noarch

#如果提示有依赖，不能卸载，请在 -e 后面加上 --nodeps 参数
```

（4）安装 JDK

1）上传 JDK。

使用 SFTP 工具将 JDK 导入到 opt 目录下面的 software 文件夹。

2）解压 JDK 到/opt/module 目录下。

```
#进入/opt/software 目录
[Hadoop@Hadoop102 opt]$ cd /opt/software/

#解压jdk 到/opt/module 目录下
[Hadoop@Hadoop102 software]$ tar -zxvf jdk-8u144-linux-x64.tar.gz -C
/opt/module/
```

3）配置 JDK 环境变量。

```
#进入JDK 所在的目录
[Hadoop@Hadoop102 software]$ cd /opt/module/jdk1.8.0_144/

#获取JDK 路径
[Hadoop@Hadoop102 jdk1.8.0_144]$ pwd
/opt/module/jdk1.8.0_144

#编辑 /etc/prpfile.d/env.sh 文件
[Hadoop@Hadoop102 jdk1.8.0_144]$ sudo vim /etc/profile.d/env.sh

#追加内容如下
##JAVA_HOME
export JAVA_HOME=/opt/module/jdk1.8.0_144
export PATH=$PATH:$JAVA_HOME/bin

#刷新环境变量
[Hadoop@Hadoop102 jdk1.8.0_144]$ source /etc/profile.d/env.sh

#检查jdk 是否安装成功，如下所示，安装成功
[Hadoop@Hadoop102 jdk1.8.0_144]$ java -version
java version "1.8.0_144"
Java(TM) SE Runtime Environment (build 1.8.0_144-b01)
Java HotSpot(TM) 64-Bit Server VM (build 25.144-b01, mixed mode)
```

（5）安装 Hadoop

1）上传 Hadoop。

使用 SFTP 工具将 Hadoop 导入到 opt 目录下面的 software 文件夹。

2）解压 Hadoop 到/opt/module 目录下。

```
#进入 /opt/software 目录
[Hadoop@Hadoop102 jdk1.8.0_144]$ cd /opt/software/

#解压 Hadoop 至 /opt/module
[Hadoop@Hadoop102 software]$ tar -zxvf Hadoop-2.7.2.tar.gz -C /opt/module/
```

3）配置 Hadoop 环境变量。

```
#进入 Hadoop 文件夹所在的目录
[Hadoop@Hadoop102 software]$ cd /opt/module/Hadoop-2.7.2/

#获取 Hadoop 路径
[Hadoop@Hadoop102 Hadoop-2.7.2]$ pwd
/opt/module/Hadoop-2.7.2
```

```
#编辑/etc/prpfile.d/env.sh 文件
[Hadoop@Hadoop102 Hadoop-2.7.2]$ sudo vim /etc/profile.d/env.sh

#追加内容如下：
##Hadoop_HOME
export Hadoop_HOME=/opt/module/Hadoop-2.7.2
export PATH=$PATH:$Hadoop_HOME/bin
export PATH=$PATH:$Hadoop_HOME/sbin

#刷新环境变量
[Hadoop@Hadoop102 Hadoop-2.7.2]$ source /etc/profile.d/env.sh

#检查 Hadoop 是否安装成功，如下所示，安装成功
[Hadoop@Hadoop102 Hadoop-2.7.2]$ Hadoop version
Hadoop 2.7.2
Subversion Unknown -r Unknown
Compiled by root on 2017-05-22T10:49Z
Compiled with protoc 2.5.0
From source with checksum d0fda26633fa762bff87ec759ebe689c
This command was run using /opt/module/Hadoop-2.7.2/share/Hadoop/common/Hadoop-common-2.7.2.jar
[Hadoop@Hadoop102 Hadoop-2.7.2]$
```

注意：后续操作 Hadoop 请使用 Hadoop 用户。

使用单台虚拟机就可以了，请继续重复 2）～4）的步骤配置另外两台虚拟机（当然，在本地也可以在配置好的虚拟机上复制两台虚拟机），记得修改主机名和 IP 地址。

（6）编写完全分布式脚本

1）编写集群分发脚本 xsync。

```
[Hadoop@Hadoop102 ~]$ mkdir bin/
[Hadoop@Hadoop102 ~]$ cd bin/
[Hadoop@Hadoop102 bin]$ touch xsync
[Hadoop@Hadoop102 bin]$ ls
xsync
[Hadoop@Hadoop102 bin]$ vim xsync
```

2）在 xsync 中编写如下代码。

```
#!/bin/bash
#1 获取输入参数个数，如果没有参数，直接退出
pcount=$#
if((pcount==0)); then
echo no args;
exit;
fi

#2 获取文件名称
p1=$1
fname=`basename $p1`
echo fname=$fname
```

```
#3 获取上级目录到绝对路径
pdir='cd -P $(dirname $p1); pwd'
echo pdir=$pdir

#4 获取当前用户名称
user='whoami'

#5 循环
for((host=103; host<105; host++)); do
        echo ------------------ Hadoop$host --------------
        rsync -rvl $pdir/$fname $user@Hadoop$host:$pdir
done
```

3）为脚本 xsync 添加执行权限。

```
[Hadoop@Hadoop102 bin]$ chmod 777 xsync
```

4）调用脚本形式：xsync 文件名称。

```
[Hadoop@Hadoop102 bin]$ xsync /home/Hadoop/bin/
```

（7）集群配置

1）分别对主机 Hadoop102，Hadoop103 和 Hadoop104 实施集群部署规划，如表 6-1 所示。

表 6-1　集群部署规划

主机名 组件名	Hadoop102	Hadoop103	Hadoop104
HDFS	NameNode DataNode	DataNode	SecondaryNameNode DataNode
YARN	NodeManager	ResourceManager NodeManager	NodeManager

2）配置集群。

配置 core-site.xml。

```xml
<!-- 指定 HDFS 中 NameNode 的地址 -->
<property>
        <name>fs.defaultFS</name>
        <value>hdfs://Hadoop102:9000</value>
</property>

<!-- 指定 Hadoop 运行时产生文件的存储目录 -->
<property>
        <name>Hadoop.tmp.dir</name>
        <value>/opt/module/Hadoop-2.7.2/data/tmp</value>
</property>
```

配置 Hadoop-env.sh。

```
[Hadoop@Hadoop102 Hadoop]$ vim Hadoop-env.sh
export JAVA_HOME=/opt/module/jdk1.8.0_144
```

配置 hdfs-site.xml。

```
[Hadoop@Hadoop102 Hadoop]$ vim hdfs-site.xml
<property>
        <name>dfs.replication</name>
        <value>3</value>
```

```
    </property>

    <!-- 指定 Hadoop 辅助名称节点主机配置 -->
    <property>
        <name>dfs.namenode.secondary.http-address</name>
        <value>Hadoop104:50090</value>
    </property>
```

配置 yarn-env.sh。

```
[Hadoop@Hadoop102 Hadoop]$ vim yarn-env.sh

export JAVA_HOME=/opt/module/jdk1.8.0_144
```

配置 yarn-site.xml。

```
[Hadoop@Hadoop102 Hadoop]$ vim yarn-site.xml
<!-- Reducer 获取数据的方式 -->
<property>
        <name>yarn.nodemanager.aux-services</name>
        <value>mapreduce_shuffle</value>
</property>

<!-- 指定 YARN 的 ResourceManager 的地址 -->
<property>
        <name>yarn.resourcemanager.hostname</name>
        <value>Hadoop103</value>
</property>
```

配置 mapred-env.sh。

```
[Hadoop@Hadoop102 Hadoop]$ vim mapred-env.sh

export JAVA_HOME=/opt/module/jdk1.8.0_144
```

配置 mapred-site.xml。

```
[Hadoop@Hadoop102 Hadoop]$ cp mapred-site.xml.template mapred-site.xml

[Hadoop@Hadoop102 Hadoop]$ vim mapred-site.xml
<!-- 指定 MR 运行在 YARN 上 -->
<property>
        <name>mapreduce.framework.name</name>
        <value>yarn</value>
</property>
```

在集群上分发配置好的 Hadoop 配置文件。

```
[Hadoop@Hadoop102 Hadoop]$ xsync /opt/module/Hadoop-2.7.2/etc/Hadoop
```

查看文件分发情况。

```
[Hadoop@Hadoop103 Hadoop-2.7.2]$ cat /opt/module/Hadoop-2.7.2/etc/Hadoop/core-site.xml
```

(8) 集群单点启动

1) 格式化 NameNode。

注意：第一次启动集群应格式化 NameNode。

```
[Hadoop@Hadoop102 Hadoop-2.7.2]$ Hadoop namenode -format
```

2）在 Hadoop102 上启动 NameNode。

```
[Hadoop@Hadoop102 Hadoop-2.7.2]$ Hadoop-daemon.sh start namenode
starting namenode, logging to /opt/module/Hadoop-2.7.2/logs/Hadoop-Hadoop-namenode-Hadoop102.out
[Hadoop@Hadoop102 Hadoop-2.7.2]$ jps
56403 Jps
56334 NameNode
```

3）在 Hadoop102、Hadoop103 以及 Hadoop104 上分别启动 DataNode。

```
[Hadoop@Hadoop102 Hadoop-2.7.2]$ Hadoop-daemon.sh start datanode
starting datanode, logging to /opt/module/Hadoop-2.7.2/logs/Hadoop-Hadoop-datanode-Hadoop102.out
[Hadoop@Hadoop102 Hadoop-2.7.2]$ jps
56450 DataNode
56525 Jps
56334 NameNode

[Hadoop@Hadoop103 Hadoop-2.7.2]$ Hadoop-daemon.sh start datanode
starting datanode, logging to /opt/module/Hadoop-2.7.2/logs/Hadoop-Hadoop-datanode-Hadoop103.out
[Hadoop@Hadoop103 Hadoop-2.7.2]$ jps
56513 Jps
56436 DataNode

[Hadoop@Hadoop104 Hadoop-2.7.2]$ Hadoop-daemon.sh start datanode
starting datanode, logging to /opt/module/Hadoop-2.7.2/logs/Hadoop-Hadoop-datanode-Hadoop104.out
[Hadoop@Hadoop104 Hadoop-2.7.2]$ jps
56386 DataNode
56471 Jps
```

（9）SSH 免密登录配置

1）生成公钥和私钥。

```
[Hadoop@Hadoop102 Hadoop-2.7.2]$ ssh-keygen -t rsa
```

按三次〈Enter〉键，就会生成两个文件 id_rsa（私钥）、id_rsa.pub（公钥）。

2）将公钥复制到要免密登录的目标机器上。

```
[Hadoop@Hadoop102 Hadoop-2.7.2]$ ssh-copy-id Hadoop102
[Hadoop@Hadoop102 Hadoop-2.7.2]$ ssh-copy-id Hadoop104
[Hadoop@Hadoop102 Hadoop-2.7.2]$ ssh-copy-id Hadoop103
```

 注意：还需要在 Hadoop103 上采用 Hadoop 账号配置 ssh 免密登录。

（10）集群配置

1）配置 slaves。

```
[Hadoop@Hadoop103 Hadoop]$ pwd
/opt/module/Hadoop-2.7.2/etc/Hadoop
```

```
[Hadoop@Hadoop103 Hadoop]$ vim slaves

#把localhost 修改为如下内容
Hadoop102
Hadoop103
Hadoop104
```

 注意：该文件添加的内容结尾不允许有空格，文件中不允许有空行。

同步所有节点配置文件。

```
[Hadoop@Hadoop103 Hadoop]$ xsync slaves
```

2）启动 HDFS。

启动之前记得先检查进程，没停止的进程先停止再启动。

```
[Hadoop@Hadoop102 Hadoop-2.7.2]$ sbin/start-dfs.sh
[Hadoop@Hadoop102 Hadoop-2.7.2]$ jps
57792 NameNode
58162 Jps
57933 DataNode

[Hadoop@Hadoop103 Hadoop]$ jps
57897 Jps
57789 DataNode

[Hadoop@Hadoop104 Hadoop]$ jps
57841 Jps
57658 DataNode
57775 SecondaryNameNode
```

3）启动 YARN。

```
[Hadoop@Hadoop103 Hadoop-2.7.2]$ sbin/start-yarn.sh
[Hadoop@Hadoop103 Hadoop-2.7.2]$ jps
58292 Jps
57963 ResourceManager
58075 NodeManager
57789 DataNode

[Hadoop@Hadoop102 Hadoop-2.7.2]$ jps
59376 Jps
59240 NodeManager
58795 NameNode
58941 DataNode

[Hadoop@Hadoop104 Hadoop]$ jps
58035 Jps
57658 DataNode
57916 NodeManager
57775 SecondaryNameNode
```

 注意：在 Hadoop103 上启动 YARN。

4）Web 端查看 SecondaryNameNode。

浏览器中输入http://Hadoop104:50090/status.html。

（11）Hadoop 常用操作命令

1）HDFS 的三种 shell 命令方式。

- Hadoop fs：适用于任何不同的文件系统，比如本地文件系统和 HDFS 文件系统。
- Hadoop dfs：只适用于 HDFS 文件系统。
- hdfs dfs：跟 Hadoop dfs 的命令作用一样，也只适用于 HDFS 文件系统。

2）查看帮助。

```
[Hadoop@Hadoop102 Hadoop-2.7.2]$ Hadoop fs
Usage: Hadoop fs [generic options]
        [-appendToFile <localsrc> ... <dst>]
        [-cat [-ignoreCrc] <src> ...]
        [-checksum <src> ...]
        [-chgrp [-R] GROUP PATH...]
        [-chmod [-R] <MODE[,MODE]... | OCTALMODE> PATH...]
        [-chown [-R] [OWNER][:[GROUP]] PATH...]
        [-copyFromLocal [-f] [-p] [-l] <localsrc> ... <dst>]
        [-copyToLocal [-p] [-ignoreCrc] [-crc] <src> ... <localdst>]
        [-count [-q] [-h] <path> ...]
        [-cp [-f] [-p | -p[topax]] <src> ... <dst>]
        [-createSnapshot <snapshotDir> [<snapshotName>]]
        [-deleteSnapshot <snapshotDir> <snapshotName>]
        [-df [-h] [<path> ...]]
        [-du [-s] [-h] <path> ...]
        [-expunge]
        [-find <path> ... <expression> ...]
        [-get [-p] [-ignoreCrc] [-crc] <src> ... <localdst>]
        [-getfacl [-R] <path>]
        [-getfattr [-R] {-n name | -d} [-e en] <path>]
        [-getmerge [-nl] <src> <localdst>]
        [-help [cmd ...]]
        [-ls [-d] [-h] [-R] [<path> ...]]
        [-mkdir [-p] <path> ...]
        [-moveFromLocal <localsrc> ... <dst>]
        [-moveToLocal <src> <localdst>]
        [-mv <src> ... <dst>]
        [-put [-f] [-p] [-l] <localsrc> ... <dst>]
        [-renameSnapshot <snapshotDir> <oldName> <newName>]
        [-rm [-f] [-r|-R] [-skipTrash] <src> ...]
        [-rmdir [--ignore-fail-on-non-empty] <dir> ...]
        [-setfacl [-R] [{-b|-k} {-m|-x <acl_spec>} <path>]|[--set <acl_spec> <path>]]
        [-setfattr {-n name [-v value] | -x name} <path>]
        [-setrep [-R] [-w] <rep> <path> ...]
        [-stat [format] <path> ...]
        [-tail [-f] <file>]
        [-test -[defsz] <path>]
```

```
            [-text [-ignoreCrc] <src> ...]
            [-touchz <path> ...]
            [-truncate [-w] <length> <path> ...]
            [-usage [cmd ...]]

        Generic options supported are
        -conf <configuration file>     specify an application configuration file
        -D <property=value>            use value for given property
        -fs <local|namenode:port>      specify a namenode
        -jt <local|resourcemanager:port>    specify a ResourceManager
        -files <comma separated list of files>    specify comma separated files
to be copied to the map reduce cluster
        -libjars <comma separated list of jars>    specify comma separated jar
files to include in the classpath.
        -archives <comma separated list of archives>    specify comma separated
archives to be unarchived on the compute machines.

        The general command line syntax is
        bin/Hadoop command [genericOptions] [commandOptions]
```

3）查看某条命令具体的作用。

```
        #格式： Hadoop fs -help 某个命令的名称
        [Hadoop@Hadoop102 Hadoop-2.7.2]$ Hadoop fs -help cp
        -cp [-f] [-p | -p[topax]] <src> ... <dst> :
          Copy files that match the file pattern <src> to a destination. When
copying
          multiple files, the destination must be a directory. Passing -p preserves
status
          [topax] (timestamps, ownership, permission, ACLs, XAttr). If -p is specified
          with no <arg>, then preserves timestamps, ownership, permission. If -pa is
          specified, then preserves permission also because ACL is a super-set of
          permission. Passing -f overwrites the destination if it already exists. raw
          namespace extended attributes are preserved if (1) they are supported (HDFS
          only) and, (2) all of the source and target pathnames are in the
/.reserved/raw
          hierarchy. raw namespace xattr preservation is determined solely by the
presence
          (or absence) of the /.reserved/raw prefix and not by the -p option.
        [Hadoop@Hadoop102 Hadoop-2.7.2]$
```

(12) 在 HDFS 文件系统上操作的具体应用

1）创建一个 input 文件夹。

```
        [Hadoop@Hadoop102 Hadoop-2.7.2]$ Hadoop fs -mkdir -p /usr/abes/input
```

2）查看 input 文件夹。

```
        [Hadoop@Hadoop102 Hadoop-2.7.2]$ hdfs dfs -ls -R /
        drwx------    - Hadoop supergroup    0 2021-04-01 15:43 /tmp
        drwx------    - Hadoop supergroup    0 2021-04-01 15:43 /tmp/Hadoop-yarn
        drwx------    - Hadoop supergroup    0 2021-04-01 15:43 /tmp/Hadoop-yarn/staging
        drwx------    - Hadoop supergroup    0 2021-04-01 15:43 /tmp/Hadoop-yarn/
```

```
staging/Hadoop
    drwx------   - Hadoop supergroup    0  2021-04-01  15:48  /tmp/Hadoop-yarn/
staging/Hadoop/.staging
    drwxr-xr-x   - Hadoop supergroup    0 2021-04-01 16:18 /usr
    drwxr-xr-x   - Hadoop supergroup    0 2021-04-01 16:18 /usr/abes
    drwxr-xr-x   - Hadoop supergroup    0 2021-04-01 16:18 /usr/abes/input
```

3）本地创建测试文件。

```
#在 Hadoop-2.7.2 目录下创建 wcinput 文件夹
#在 input 目录下创建 wc.input 文件
#编辑 wc.input 文件
内容如下：
Hadoop hdfs
Hadoop mapreduce
abes
abes
```

4）上传测试文件到 input。

上传 wc.input 至 hdfs 文件系统下的/usr/abes/input。

```
[Hadoop@Hadoop102 Hadoop-2.7.2]$ hdfs dfs -put wcinput/wc.input /usr/abes/input
```

查看是否上传成功。

```
[Hadoop@Hadoop102 Hadoop-2.7.2]$ hdfs dfs -ls /usr/abes/input
Found 1 items
-rw-r--r--  1 Hadoop supergroup    39 2021-04-01 16:43 /usr/abes/input/wc.input
[Hadoop@Hadoop102 Hadoop-2.7.2]$
```

5）使用 MapReduce 程序运行测试文件。

使用 Mapreduce 运行 4.2 节已经创建的文件，wordcount 为词频统计。

```
[Hadoop@Hadoop102 Hadoop-2.7.2]$ Hadoop jar share/Hadoop/mapreduce/Hadoop-mapreduce-examples-2.7.2.jar wordcount /usr/abes/input /usr/abes/output
```

6）查看输出结果。

```
[Hadoop@Hadoop102 Hadoop-2.7.2]$ hdfs dfs -cat /usr/abes/output/*
abes     2
Hadoop   2
hdfs     1
mapreduce    1
```

在浏览器中查看输出结果，如图 6-6 所示。

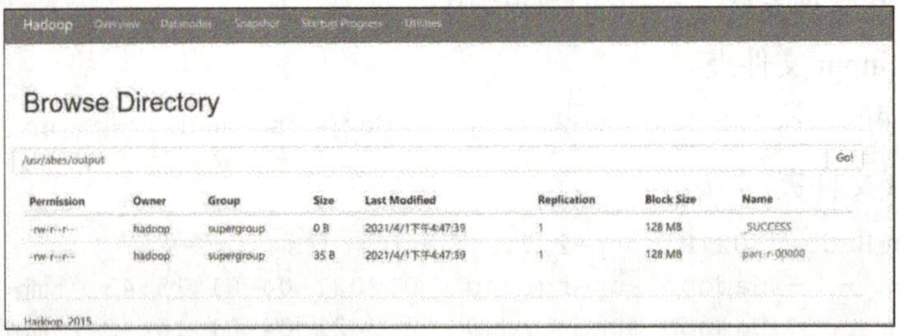

图 6-6 查看 mapreduce 输出结果

7）下载输出结果。

```
[Hadoop@Hadoop102 Hadoop-2.7.2]$ hdfs dfs -get /usr/abes/output/part-r-00000 ./wcinput/
[Hadoop@Hadoop102 Hadoop-2.7.2]$ ls wcinput/
part-r-00000  wc.input
```

8）删除输出结果。

```
[Hadoop@Hadoop102 Hadoop-2.7.2]$ hdfs dfs -rm -r /usr/abes/output
21/04/01 17:24:47 INFO fs.TrashPolicyDefault: Namenode trash configuration: Deletion interval = 0 minutes, Emptier interval = 0 minutes.
Deleted /usr/abes/output
```

6.2 Hadoop 生态组件

6.2.1 Hadoop 生态圈简介

除了 Hadoop 的三大核心组件之外，Hadoop 生态圈中常用组件包括：Hive、Spark、HBase、Kafka、Flume、Sqoop 和 Zookeeper 等。这些组件分属于不同的功能层级，共同组成了一个功能完整且高效的 Hadoop 生态圈，如图 6-7 所示。

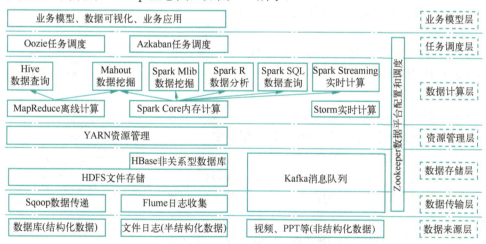

图 6-7 Hadoop 生态圈架构图

从图中可以看出，Hadoop 生态圈架构中包含了内容丰富且功能强大的组件集合。从功能上看，Hadoop 生态圈架构可以分为多个层级，具体如下。

- 数据来源层能够涵盖结构化数据（数据库）、半结构化数据（文件日志）和非结构化数据（视频、音频和 PPT 等）。
- 数据传输层能够充分发挥集群网络优势，高效地并发和并行传输数据。
- 数据存储层不仅能够充分发挥集群强大的整体存储能力，还能通过数据冗余的方式有效提高数据安全性，以及数据读写效率。
- 资源管理层能够根据集群各节点资源（计算资源、存储资源和网络资源）使用情况合理、平衡地分配和安排各节点在整个数据处理过程中的任务。
- 数据计算层能够根据具体的业务计算需求，提供离线计算和实时计算。一般对于规模和体量较大的历史业务数据采用离线计算，对于不断修改和新增的流式数据采用实时计算。离线计算适合利用非业务处理高峰时间等对实时性要求不高的业务，实时计算适合对时间和效率要求较高的业务。
- 任务调度层能够根据 Hadoop 大数据平台系统当前的任务数量和状态，按照一定的调度算法对任务进行合理调度，从而优化和平衡系统整体的运行效率。
- 业务模型层能够提供直观、简洁、精准的数据可视化呈现和应用效果，帮助用户提升业务的分析和处理能力。

6.2.2 Hive 的安装、部署和应用

（1）Hive 简介

Hive 是一套基于 Hadoop 的数据仓库工具。它能够实现 Hadoop 中数据的保存、检索和分析。同时，它还能够将 HQL 语言通过转化之后，在 MapReduce 中执行。这使得 MapReduce 开发人员能够通过 SQL 语言进行一些 MapReduce 自身无法实现的任务。

（2）Hive 的特点

1）可扩展性。

由于 Hive 是建立在 Hadoop 之上，因此有与 Hadoop 集群一样的扩展性。Hive 可以在不用重新启动服务的前提下实现集群规模的自由扩展。

2）可延展性。

Hive 可以通过编写更加灵活多样的 HQL 语言实现比 MapReduce 更丰富的函数。

3）容错性。

Hive 可以使用 Hadoop 集群的超强冗余性实现数据的容错性。Hive 本身的元数据则存放在 MySQL 数据库中。

（3）Hive 的组织架构（如图 6-8 所示）

1）访问 Hive 的接口：CLI shell、JDBC/ODBC、WebGUI。

2）Hive 元数据 Metastore 的存储位置：存储在 MySQL 数据库。

3）Hive 驱动器（Driver）：编译器（Compiler）、优化器（Optimizer）和执行器（Executor）等。

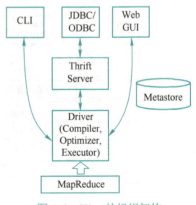

图 6-8　Hive 的组织架构

4）Hive 后台守护进程 Thrift Server：一直等待来自访问接口或驱动器的数据。

（4）下载地址

1）Hive 官网地址为：http://hive.apache.org/。

2）文档查看地址为：https://cwiki.apache.org/confluence/display/Hive/GettingStarted。

3）下载地址为：http://archive.apache.org/dist/hive/。

4）Github 地址为https://github.com/apache/hive。

（5）Hive 安装部署

1）把 Hive 上传到 Linux。

2）解压至/opt/module。

```
[Hadoop@Hadoop102 software]$ tar -zxvf apache-hive-1.2.1-bin.tar.gz -C /opt/module/
```

3）改名为 hive。

```
[Hadoop@Hadoop102 software]$ cd /opt/module/
[Hadoop@Hadoop102 module]$ ls
apache-hive-1.2.1-bin  Hadoop-2.7.2  jdk1.8.0_144
[Hadoop@Hadoop102 module]$ mv apache-hive-1.2.1-bin/ hive
[Hadoop@Hadoop102 module]$ ls
Hadoop-2.7.2  hive  jdk1.8.0_144
[Hadoop@Hadoop102 module]$
```

4）修改 conf 目录下的 hive-env.sh.template 为 hive-env.sh。

```
[Hadoop@Hadoop102 module]$ cd hive/
[Hadoop@Hadoop102 hive]$ ls
bin  conf  examples  hcatalog  lib  LICENSE  NOTICE  README.txt  RELEASE_NOTES.txt  scripts
[Hadoop@Hadoop102 hive]$ cd conf/
[Hadoop@Hadoop102 conf]$ ls
beeline-log4j.properties.template  hive-env.sh.template  hive-log4j.properties.template
hive-default.xml.template  hive-exec-log4j.properties.template  ivysettings.xml
[Hadoop@Hadoop102 conf]$ mv hive-env.sh.template hive-e
hive-env.sh.template               hive-exec-log4j.properties.template
[Hadoop@Hadoop102 conf]$ mv hive-env.sh.template hive-env.sh
[Hadoop@Hadoop102 conf]$
```

5）配置 hive-env.sh。

配置 Hadoop_HOME 路径。

```
# Set Hadoop_HOME to point to a specific Hadoop install directory
export Hadoop_HOME=/opt/module/Hadoop-2.7.2
```

配置 HIVE_CONF_DIR 路径。

```
# Hive Configuration Directory can be controlled by:
export HIVE_CONF_DIR=/opt/module/hive/conf
```

（6）启动 Hive

1）必须先启动 hdfs 和 yarn。

```
[Hadoop@Hadoop102 Hadoop-2.7.2]$ start-dfs.sh
Starting namenodes on [Hadoop102]
Hadoop102: starting namenode, logging to /opt/module/Hadoop-2.7.2/logs/Hadoop-Hadoop-namenode-Hadoop102.out
Hadoop103: starting datanode, logging to /opt/module/Hadoop-2.7.2/logs/Hadoop-Hadoop-datanode-Hadoop103.out
Hadoop102: starting datanode, logging to /opt/module/Hadoop-2.7.2/logs/Hadoop-Hadoop-datanode-Hadoop102.out
Hadoop104: starting datanode, logging to /opt/module/Hadoop-2.7.2/logs/Hadoop-Hadoop-datanode-Hadoop104.out
Starting secondary namenodes [Hadoop104]
```

	Hadoop104: starting secondarynamenode, logging to /opt/module/Hadoop-2.7.2/logs/Hadoop-Hadoop-secondarynamenode-Hadoop104.out
	[Hadoop@Hadoop102 Hadoop-2.7.2]$ jps
	71698 Jps
	71477 DataNode
	71326 NameNode

	[Hadoop@Hadoop103 Hadoop-2.7.2]$ start-yarn.sh
	starting yarn daemons
	starting resourcemanager, logging to /opt/module/Hadoop-2.7.2/logs/yarn-Hadoop-resourcemanager-Hadoop103.out
	Hadoop104: starting nodemanager, logging to /opt/module/Hadoop-2.7.2/logs/yarn-Hadoop-nodemanager-Hadoop104.out
	Hadoop102: starting nodemanager, logging to /opt/module/Hadoop-2.7.2/logs/yarn-Hadoop-nodemanager-Hadoop102.out
	Hadoop103: starting nodemanager, logging to /opt/module/Hadoop-2.7.2/logs/yarn-Hadoop-nodemanager-Hadoop103.out
	[Hadoop@Hadoop103 Hadoop-2.7.2]$ jps
	67929 DataNode
	68392 Jps
	68058 ResourceManager
	68175 NodeManager

2）启动 hive。

	[Hadoop@Hadoop102 Hadoop-2.7.2]$ cd ..
	[Hadoop@Hadoop102 module]$ cd hive/
	[Hadoop@Hadoop102 hive]$ bin/hive

	Logging initialized using configuration in jar:file:/opt/module/hive/lib/hive-common-1.2.1.jar!/hive-log4j.properties
	hive>

（7）Hive 基本操作

1）查看数据库。

	hive> show databases;
	OK
	default
	Time taken: 0.861 seconds, Fetched: 1 row(s)

2）打开默认数据库。

	hive> use default;
	OK
	Time taken: 0.018 seconds

3）显示 default 数据库中的表。

	hive> show tables;
	OK
	Time taken: 0.026 seconds

4）创建一张表。

	hive> create table student(id int,name string);

OK
Time taken: 0.446 seconds

5）显示数据库中有几张表。

```
hive> show tables;
OK
student
Time taken: 0.021 seconds, Fetched: 1 row(s)
```

6）查看表的结构。

```
hive> desc student;
OK
id                      int
name                    string
Time taken: 0.351 seconds, Fetched: 2 row(s)
```

7）向表中插入数据。

```
hive> insert into student values(1001,"AA");
Query ID = Hadoop_20210402142949_a9fdef52-adee-4879-8287-c58fa73c29c0
Total jobs = 3
Launching Job 1 out of 3
Number of reduce tasks is set to 0 since there's no reduce operator
Starting Job = job_1617344159555_0001, Tracking URL = http://Hadoop103:8088/proxy/application_1617344159555_0001/
Kill Command = /opt/module/Hadoop-2.7.2/bin/Hadoop job  -kill job_1617344159555_0001
Hadoop job information for Stage-1: number of mappers: 1; number of reducers: 0
2021-04-02 14:30:00,710 Stage-1 map = 0%,  reduce = 0%
2021-04-02 14:30:09,516 Stage-1 map = 100%,  reduce = 0%, Cumulative CPU 2.78 sec
MapReduce Total cumulative CPU time: 2 seconds 780 msec
Ended Job = job_1617344159555_0001
Stage-4 is selected by condition resolver.
Stage-3 is filtered out by condition resolver.
Stage-5 is filtered out by condition resolver.
Moving data to: hdfs://Hadoop102:9000/user/hive/warehouse/student/.hive-staging_hive_2021-04-02_14-29-49_035_2372269803035345041-1/-ext-10000
Loading data to table default.student
Table default.student stats: [numFiles=1, numRows=1, totalSize=8, rawDataSize=7]
MapReduce Jobs Launched:
Stage-Stage-1: Map: 1   Cumulative CPU: 2.78 sec   HDFS Read: 3563 HDFS Write: 79 SUCCESS
Total MapReduce CPU Time Spent: 2 seconds 780 msec
OK
Time taken: 22.984 seconds
```

8）查询表中的数据。

```
hive> select * from student;
OK
1001    AA
Time taken: 0.167 seconds, Fetched: 1 row(s)
```

9）退出 Hive。

```
hive> quit;
[Hadoop@Hadoop102 hive]$
```

（8）将本地文件导入 Hive 应用

1）数据准备。

在/opt/module/目录下创建一个 datas 的目录，并在其下创建一个 student.txt 文件，向该文件写入数据。

```
[Hadoop@Hadoop102 hive]$ cd ..
[Hadoop@Hadoop102 module]$ mkdir datas
[Hadoop@Hadoop102 module]$ cd datas/
[Hadoop@Hadoop102 datas]$ touch student.txt
[Hadoop@Hadoop102 datas]$ vim student.txt

1001 hive
1002 mzmz
1003 abes
```

 注意：列与列之间使用一个空格键进行分割。

2）导入数据。

```
#启动 hive
[Hadoop@Hadoop102 hive]$ bin/hive

Logging initialized using configuration in jar:file:/opt/module/hive/lib/hive-common-1.2.1.jar!/hive-log4j.properties
    hive>

#显示数据库
hive> show databases;

#使用 defalut 数据库
hive> use default;

#显示 defalut 数据库中的表
hive> show tables;

#删除已经创建的 student 表，没有就不用管了
hive> drop table student;

#创建 student 表，并声明文件分隔符'\t'
hive> create table student(id int, name string) ROW FORMAT DELIMITED FIELDS TERMINATED
    > BY '\t';

#加载/opt/module/datas/student.txt 文件到 student 表中
hive> load data local inpath '/opt/module/datas/student.txt' into table student;
Loading data to table default.student
```

```
Table default.student stats: [numFiles=1, totalSize=39]
OK
Time taken: 0.835 seconds

#查看结果
hive> select * from student;
OK
1001    hive
1002    mzmz
1003    abes
Time taken: 0.281 seconds, Fetched: 3 row(s)
```

(9)MySQL 的安装部署

1)安装的缘由。

Hive 默认使用 Metastore 自带的 derby 数据库,但该数据库不能很好地适应真实的应用场景,这里通过代码进行说明。这里使用基于 derby 数据库启动第一个 hive 客户端窗口:

```
[Hadoop@Hadoop102 hive]$ bin/hive
Logging initialized using configuration in jar:file:/opt/module/hive/lib/hive-common-1.2.1.jar!/hive-log4j.properties
hive>
```

显示启动成功了,这时尝试新开一个 Hive 窗口。

```
[Hadoop@Hadoop102 hive]$ bin/hive
Logging initialized using configuration in jar:file:/opt/module/hive/lib/hive-common-1.2.1.jar!/hive-log4j.properties
Exception in thread "main" java.lang.RuntimeException: java.lang.RuntimeException:
   Unable to instantiate org.apache.Hadoop.hive.ql.metadata.SessionHiveMetaStoreClient
        at org.apache.Hadoop.hive.ql.session.SessionState.start(SessionState.java:522)
        at org.apache.Hadoop.hive.cli.CliDriver.run(CliDriver.java:677)
        at org.apache.Hadoop.hive.cli.CliDriver.main(CliDriver.java:621)
        at sun.reflect.NativeMethodAccessorImpl.invoke0(Native Method)
        at  sun.reflect.NativeMethodAccessorImpl.invoke(NativeMethodAccessorImpl.java:62)
        at sun.reflect.DelegatingMethodAccessorImpl.invoke(DelegatingMethodAccessorImpl.java:43)
        at java.lang.reflect.Method.invoke(Method.java:498)
        at org.apache.Hadoop.util.RunJar.run(RunJar.java:221)
        at org.apache.Hadoop.util.RunJar.main(RunJar.java:136)
Caused by: java.lang.RuntimeException: Unable to instantiate org.apache.Hadoop.hive.ql.metadata.SessionHiveMetaStoreClient
        at org.apache.Hadoop.hive.metastore.MetaStoreUtils.newInstance(MetaStoreUtils.java:1523)
        at org.apache.Hadoop.hive.metastore.RetryingMetaStoreClient.<init>(RetryingMetaStoreClient.java:86)
        at org.apache.Hadoop.hive.metastore.RetryingMetaStoreClient.getProxy(RetryingMetaStoreClient.java:132)
        at org.apache.Hadoop.hive.metastore.RetryingMetaStoreClient.getProxy(RetryingMetaStoreClient.java:104)
        at org.apache.Hadoop.hive.ql.metadata.Hive.createMetaStoreClient(Hive.java:3005)
```

```
        at org.apache.Hadoop.hive.ql.metadata.Hive.getMSC(Hive.java:3024)
        at org.apache.Hadoop.hive.ql.session.SessionState.start(SessionState.java:503)
        ... 8 more
Caused by: java.lang.reflect.InvocationTargetException
        at sun.reflect.NativeConstructorAccessorImpl.newInstance0(Native Method)
        at sun.reflect.NativeConstructorAccessorImpl.newInstance(NativeConstructorAccessorImpl.java:62)
        at sun.reflect.DelegatingConstructorAccessorImpl.newInstance(DelegatingConstructorAccessorImpl.java:45)
        at java.lang.reflect.Constructor.newInstance(Constructor.java:423)
        at org.apache.Hadoop.hive.metastore.MetaStoreUtils.newInstance(MetaStoreUtils.java:1521)
        ... 14 more
```

发现它报错了！在 derby 中，只能使用一个客户端连接，第二个窗口自然就不能连接了。但是在生产中不可能只一个客户端使用 Hive，所以使用 MySQL 存储 Metastore 来解决这个问题。

2）安装准备。

```
#把 MySQL 安装包发送到 Linux 下的/opt/software
#查询 MySQL 是否安装，如果安装了，需要卸载
[Hadoop@Hadoop102 hive]$ rpm -qa | grep mariadb
mariadb-libs-5.5.64-1.el7.x86_64
[Hadoop@Hadoop102 hive]$ sudo rpm -e --nodeps mariadb-libs

#解压 MySQL
[Hadoop@Hadoop102 hive]$ cd /opt/software/
[Hadoop@Hadoop102 software]$ unzip mysql-libs.zip
Archive:  mysql-libs.zip
   creating: mysql-libs/
  inflating: mysql-libs/MySQL-client-5.6.24-1.el6.x86_64.rpm
  inflating: mysql-libs/mysql-connector-java-5.1.27.tar.gz
  inflating: mysql-libs/MySQL-server-5.6.24-1.el6.x86_64.rpm
[Hadoop@Hadoop102 software]$ ls
apache-hive-1.2.1-bin.tar.gz  Hadoop-2.7.2.tar.gz  jdk-8u144-linux-x64.tar.gz  mysql-libs  mysql-libs.zip

#进入 mysql-libs 目录下
[Hadoop@Hadoop102 software]$ cd mysql-libs/
[Hadoop@Hadoop102 mysql-libs]$ ls
MySQL-client-5.6.24-1.el6.x86_64.rpm  mysql-connector-java-5.1.27.tar.gz
MySQL-server-5.6.24-1.el6.x86_64.rpm

#安装服务器端和客户端
[Hadoop@Hadoop102 mysql-libs]$ rpm -ivh MySQL-server-5.6.24-1.el6.x86_64.rpm
[root@Hadoop102 mysql-libs]# rpm -ivh MySQL-client-5.6.24-1.el6.x86_64.rpm
```

3）启动 MySQL 服务。

```
#切换至 root 用户查看产生的随机密码
[Hadoop@Hadoop102 mysql-libs]$ su
密码：
[root@Hadoop102 mysql-libs]# cat /root/.mysql_secret
```

```
# The random password set for the root user at Fri Apr  2 17:00:30 2021
(local time): KOzi3kIBJT382yeu
```

```
#记住临时登录密码: KOzi3kIBJT382yeu
```

```
#切换至 Hadoop,启动 MySQL
[Hadoop@Hadoop102 mysql-libs]$ sudo systemctl start mysql
```

```
#查看 MySQL 的状态
[Hadoop@Hadoop102 mysql-libs]$ sudo systemctl status mysql
 mysql.service - LSB: start and stop MySQL
   Loaded: loaded (/etc/rc.d/init.d/mysql; bad; vendor preset: disabled)
   Active: active (running) since 五 2021-04-02 17:19:25 CST; 57s ago
     Docs: man:systemd-sysv-generator(8)
   Process: 77114 ExecStop=/etc/rc.d/init.d/mysql stop (code=exited, status=0/SUCCESS)
   Process: 77154 ExecStart=/etc/rc.d/init.d/mysql start (code=exited, status=0/SUCCESS)
    Tasks: 22
   CGroup: /system.slice/mysql.service
           ├─77161 /bin/sh /usr/bin/mysqld_safe --datadir=/var/lib/mysql
--pid-file=/var/lib/mysql/Hadoop102.pid
           └─77264 /usr/sbin/mysqld --basedir=/usr --datadir=/var/lib/mysql --plugin-dir=/usr/lib64/mysql/plugin --user=mys...

4月 02 17:19:24 Hadoop102 systemd[1]: Starting LSB: start and stop MySQL...
4月 02 17:19:25 Hadoop102 mysql[77154]: Starting MySQL. SUCCESS!
4月 02 17:19:25 Hadoop102 systemd[1]: Started LSB: start and stop MySQL.
```

4)登录 MySQL。

```
#登录 MySQL 客户端
[Hadoop@Hadoop102 mysql-libs]$ mysql -uroot -pKOzi3kIBJT382yeu
   #有警告提示是因为在命令行输入了密码,-p 后不输入密码直接按〈Enter〉键,再输入密码就没
有警告了
   #修改 root 用户密码,这是必须执行的操作
mysql> set password = password('123456');
```

```
#退出
mysql> quit;
```

5)配置 MySQL 库中的 user 表。

使用 root 用户+密码在任何主机上都能登录 MySQL 数据库,需要修改 user 表。

```
#登录 MySQL
[Hadoop@Hadoop102 mysql-libs]$ mysql -uroot -p123456
```

```
#显示数据库
mysql> show databases;
```

```
#使用 MySQL 数据库
mysql> use mysql;
```

```
#展示MySQL数据库中的所有表
mysql> show tables;

#展示user表的结构
mysql> desc user;

#查询user表
mysql> select user,host from user;
+------+-----------+
| user | host      |
+------+-----------+
| root | 127.0.0.1 |
| root | ::1       |
| root | Hadoop102 |
| root | localhost |
+------+-----------+
4 rows in set (0.01 sec)

#修改user表，把Host表内容修改为%
mysql> update user set host='%' where host='localhost';

#删除root用户的其他host
mysql> delete from user where host='127.0.0.1';
mysql> delete from user where host='::1';
mysql> delete from user where host='Hadoop102';

#查询user表
mysql> select user,host from user;
+------+------+
| user | host |
+------+------+
| root | %    |
+------+------+
1 row in set (0.00 sec)

#刷新权限
mysql> flush privileges;

#退出
mysql> quit;
```

（10）配置Hives元数据到MySQL

1）复制驱动。

```
#解压mysql-connector-java-5.1.27.tar.gz
[Hadoop@Hadoop102 mysql-libs]$ tar -zxvf mysql-connector-java-5.1.27.tar.gz

#复制mysql-connector-java-5.1.27-bin.jar至/opt/module/hive/lib
[Hadoop@Hadoop102 mysql-libs]$ cd mysql-connector-java-5.1.27/
[Hadoop@Hadoop102 mysql-connector-java-5.1.27]$ cp mysql-connector-java-5.1.27-bin.jar /opt/module/hive/lib/
```

2）配置 Metastore 到 MySQL。

```
#在/opt/module/hive/conf 目录下创建一个 hive-site.xml
[Hadoop@Hadoop102 mysql-connector-java-5.1.27]$ cd /opt/module/hive/conf/
[Hadoop@Hadoop102 conf]$ vim hive-site.xml

#写入以下数据
<?xml version="1.0"?>
<?xml-stylesheet type="text/xsl" href="configuration.xsl"?>
<configuration>
    <property>
      <name>javax.jdo.option.ConnectionURL</name>
      <value>jdbc:mysql://Hadoop102:3306/metastore?createDatabaseIfNot-
Exist=true</value>
      <description>JDBC connect string for a JDBC metastore</description>
    </property>

    <property>
      <name>javax.jdo.option.ConnectionDriverName</name>
      <value>com.mysql.jdbc.Driver</value>
      <description>Driver class name for a JDBC metastore</description>
    </property>

    <property>
      <name>javax.jdo.option.ConnectionUserName</name>
      <value>root</value>
      <description>username to use against metastore database</description>
    </property>

    <property>
      <name>javax.jdo.option.ConnectionPassword</name>
      <value>123456</value>
      <description>password to use against metastore database</description>
    </property>
</configuration>

#配置完毕，如果启动 Hive 异常，可以重新启动虚拟机（重启后，别忘了启动 Hadoop 集群）
```

（11）多窗口打开测试（如图 6-9 所示）

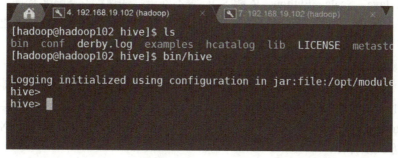

图 6-9　启动 Hive 的窗口

1)启动 MySQL。

```
[Hadoop@Hadoop102 hive]$ mysql -uroot -p123456
```

2)多开几个窗口打开 Hive。

3)查看 MySQL 中的数据库。

```
mysql> show databases;
+--------------------+
| Database           |
+--------------------+
| information_schema |
| metastore          |
| mysql              |
| performance_schema |
| test               |
+--------------------+
5 rows in set (0.00 sec)
```

可以看到 MySQL 中多了 metastores 数据库。

6.2.3 Spark 的安装、部署和应用

(1) Spark 简介

Spark 是用于大规模数据处理的统一分析引擎。与 MapReduce 使用基于 HDFS 的磁盘读写方式不同,Spark 使用基于内存的分布式并行计算框架,能够将 Hadoop 集群中运行的应用程序在内存中运行的速度提升 100 倍,甚至能够将应用程序在磁盘上的运行速度提升 10 倍。Spark 主要用于提高计算速度以满足不同的业务需求,而不是取代 Hadoop 及其 MapReduce 计算框架。Spark 不仅可以独立运行(使用 standalone 模式),还可以运行在 YARN 管理的集群中。Spark 能够很好地集成在 Hadoop 生态环境之中,并有效提高包括 Hive 等的数据分析和处理的能力。

(2) Spark 的特点

1)运行速度快。

Spark 使用基于高速缓存的分布式实时计算框架。与 MapReduce 不断重复使用磁盘输入输出保存计算结果相比,Spark 的最大优势在于能够高效地使用高速缓存进行迭代重复计算。只有在内存不足的情况下,Spark 才会使用磁盘输入和输出。

2)简单易用。

可以使用 Java、Scala、Python、R 和 SQL 等不同语言快速编写 Spark 程序,这极大地提升了 Spark 的普及范围。同时,Spark 自带 80 多个高级操作符,能够更容易地创建 Spark 并行应用程序,并能够与 Java、Scala、Python、R 和 SQL shell 交互。

3)普适通用。

Spark 是一个分布式计算框架,不仅能够实现 MapReduce 的场景功能,更能在更多业务场景中崭露头角。Spark 既可以实现离线计算的 Spark SQL 模块,也有可以实现实时计算的 Spark Streaming 模块,Spark 还封装了常用的机器学习库 Mlib 和图计算库 GraphX。

4)运行方式多样性。

Spark 有多种运行方式:Local、StandAlone 和 YARN 等。Local 模式用于测试。在 StandAlone 模式下用户可以在 Hadoop 集群的一组或全部机器中静态分配资源,与 Hadoop MR

同时运行,用户可以在 HDFS 上运行专属的 Spark 任务。在 YARN 模式中,Hadoop 用户可以简单地把 Spark 运行在 YARN 中,像其他运行在 Spark 上层的模块一样充分利用 Spark 的强大计算能力。

(3) Spark 的组织架构(如图 6-10 所示)

1) Spark SQL 模块:允许通过 SQL 的方式实现 Spark 的离线计算。

2) Spark Streaming 模块:允许通过 Streaming 的方式实现实时计算。

3) 机器学习库 MLlib:提供了常用的机器学习算法库。

图 6-10 Spark 的组织架构

4) 图计算库 GraphX:一个分布式图处理框架,基于 Spark 提供对图计算和图挖掘简洁易用且丰富的接口,极大地方便了对分布式图处理的需求。

(4) Spark 安装地址

- 官网地址为 http://spark.apache.org/。
- 文档查看地址为 https://spark.apache.org/docs/。
- 下载地址为 https://spark.apache.org/downloads.html。

(5) Local 模式

Local 模式就是运行在一台计算机上的模式,通常用于在本机上练习和测试,可以通过以下几种方式设置 Master。

- Local:所有计算都运行在一个线程中,没有任何并行计算,通常在本机执行一些测试代码和练习的模式。
- local[K]:指定使用几个线程来运行计算,比如 local[4]就是运行 4 个 Worker 线程。通常计算机的 CPU 有几个核心(Core)就指定几个线程,最大化利用 CPU 的计算能力。
- local[*]:这种模式直接帮用户按照 CPU 最多核心(Cores)来设置线程数。

1) 安装使用 Spark。

```
[Hadoop@Hadoop102 software]$ tar -zxvf spark-2.4.7-bin-Hadoop2.7.tgz -C /opt/module/
#改名
[Hadoop@Hadoop102 module]$ mv spark-2.4.7-bin-Hadoop2.7/ spark
#配置环境变量
[Hadoop@Hadoop102 spark]$ sudo vim /etc/profile.d/env.sh
#SPARK_HOME
export SPARK_HOME=/opt/module/spark
export PATH=$PATH:$SPARK_HOME/bin
export PATH=$PATH:$SPARK_HOME/sbin
[Hadoop@Hadoop102 spark]$ source /etc/profile.d/env.sh
```

2) 在 Local 模式下,使用 Spark 基础命令调用程序自带的圆周率 jar 包。

基本语法如下。

```
[Hadoop@Hadoop102 spark]$ bin/spark-submit --class org.apache.spark.examples.SparkPi --executor-memory 1G --total-executor-cores 2 ./examples/jars/spark-examples_2.11-2.4.7.jar 100
```

参数内容如下。

```
bin/spark-submit \
```

```
--class <main-class>
--master <master-url> \
--deploy-mode <deploy-mode> \
--conf <key>=<value> \
... # other options
<application-jar> \
[application-arguments]
```

参数说明如下。

--master：指定 Master 的地址
--class：应用的启动类（如 org.apache.spark.examples.SparkPi）
--deploy-mode：是否发布自己的驱动到 worker 节点(cluster)或者作为一个本地客户端(client)（default：client）
--conf：任意的 Spark 配置属性，格式 key=value。如果值包含空格，可以加引号 "key=value"
application-jar：打包好的应用 jar 包含依赖。这个 URL 在集群中全局可见。 比如 hdfs:// 共享存储系统， 如果是 file:// path， 那么所有的节点的 path 都包含同样的 jar
application-arguments：传给 main()方法的参数
--executor-memory 1G：指定每个 executor 可用内存为 1G
--total-executor-cores 2：指定总的 executor 使用的 cup 核数为两个

结果如图 6-11 所示。

（6）spark-shell 的简单使用

Spark 使用 Scala 语言开发，shell 界面也是使用 Scala。

```
[Hadoop@Hadoop102 spark]$ bin/spark-shell
21/04/07 11:22:07 WARN NativeCodeLoader: Unable to load native-Hadoop library for your platform... using builtin-java classes where applicable
Using Spark's default log4j profile: org/apache/spark/log4j-defaults.properties
Setting default log level to "WARN".
To adjust logging level use sc.setLogLevel(newLevel). For SparkR, use setLogLevel(newLevel).
Spark context Web UI available at http://Hadoop102:4040
Spark context available as 'sc' (master = local[*], app id = local-1617765736068).
Spark session available as 'spark'.
Welcome to
      ____              __
     / __/__  ___ _____/ /__
    _\ \/ _ \/ _ `/ __/  '_/
   /___/ .__/\_,_/_/ /_/\_\   version 2.4.7
      /_/

Using Scala version 2.11.12 (Java HotSpot(TM) 64-Bit Server VM, Java 1.8.0_144)
Type in expressions to have them evaluated.
Type :help for more information.

scala>

#输出 hello world
scala> println("Hello, world!")
Hello, world!
```

```
# 简单计算
scala> 1+1
res2: Int = 2

scala> 100*100
res3: Int = 10000
```

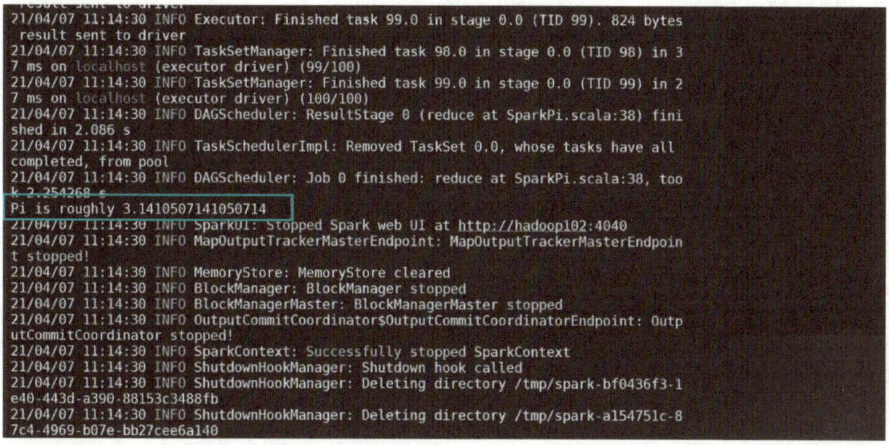

图 6-11　Pi 计算结果

（7）Standalone 模式

1）安装使用。

进入 Spark 安装目录下的 conf 文件夹。

```
[Hadoop@Hadoop102 spark]$ cd conf/
```

修改配置文件名称。

```
[Hadoop@Hadoop102 conf]$ mv spark-env.sh.template spark-env.sh
[Hadoop@Hadoop102 conf]$ mv slaves.template slaves
```

修改 slave 文件，添加 work 节点。

```
Hadoop102
Hadoop103
Hadoop104
```

修改 spark-env.sh 文件，添加如下配置。

```
[Hadoop@Hadoop102 conf]$ vim spark-env.sh
SPARK_MASTER_HOST=Hadoop102
SPARK_MASTER_PORT=7077
```

分发 Spark 包。

```
[Hadoop@Hadoop102 module]$ xsync spark/
```

2）使用 start-all.sh 脚本启动 Hadoop 平台。

```
[Hadoop@Hadoop102 spark]$ sbin/start-all.sh
[Hadoop@Hadoop102 spark]$ jps
8448 Worker
8592 Jps
```

```
109701 NameNode
109836 DataNode

[Hadoop@Hadoop103 ~]$ jps
120657 Worker
120727 Jps
114198 DataNode

[Hadoop@Hadoop104 spark]$ jps
119571 DataNode
119677 SecondaryNameNode
126860 Jps
126734 Worker
```

 注意：如果遇到"JAVA_HOME not set"异常，可以在 sbin 目录下的 spark-config.sh 文件中加入如下配置：

```
export JAVA_HOME=/opt/module/jdk1.8.0_144
```

3）使用 Spark 基础命令调用程序自带的圆周率 jar 包。

```
[Hadoop@Hadoop104 spark]$
bin/spark-submit --class org.apache.spark.examples.SparkPi --master spark://Hadoop102:7077 --executor-memory 1G --total-executor-cores 2 ./examples/jars/spark-examples_2.11-2.4.7.jar 100
```

（8）YARN 模式

1）安装使用。

修改 Hadoop 配置文件 yarn-site.xml，添加如下内容。

```
[Hadoop@Hadoop102 Hadoop]$ pwd
/opt/module/Hadoop-2.7.2/etc/Hadoop
[Hadoop@Hadoop102 Hadoop]$ vim yarn-site.xml

    <!--是否启动一个线程检查每个任务正使用的物理内存量，如果任务超出分配值，则直接将其关闭，默认是 true -->
    <property>
        <name>yarn.nodemanager.pmem-check-enabled</name>
        <value>false</value>
    </property>
    <!--是否启动一个线程检查每个任务正使用的虚拟内存量，如果任务超出分配值，则直接将其关闭，默认是 true -->
    <property>
          <name>yarn.nodemanager.vmem-check-enabled</name>
          <value>false</value>
    </property>
修改 spark-env.sh，添加如下配置
[Hadoop@Hadoop102 spark]$ vim conf/spark-env.sh
YARN_CONF_DIR=/opt/module/Hadoop-2.7.2/etc/Hadoop
分发配置文件
[Hadoop@Hadoop102 spark]$ xsync /opt/module/Hadoop-2.7.2/etc/Hadoop/yarn-site.xml
```

2)执行一个程序。

在提交任务之前需启动 HDFS 及 YARN 集群。

```
[Hadoop@Hadoop102 spark]$
bin/spark-submit --class org.apache.spark.examples.SparkPi --master yarn
--deploy-mode client ./examples/jars/spark-examples_2.11-2.4.7.jar 100
```

6.2.4 HBase 的安装、部署和应用

(1) HBase 简介

HBase 是一个非常适合处理非规范性数据、分布式、可扩展、面向列的开源关系型分布式数据库。它可以部署在大规模廉价集群中开展存储功能。它将 HDFS 作为底层文件保存系统。同时，使用 MapReduce 来处理大数据平台中的大规模数据。与 Hive 非常适合做大数据的离线查询分析相比，HBase 非常适合用来进行大数据的实时查询。

(2) HBase 的特点

1) 高可靠性。

HBase 基于 Hadoop 的 HDFS 分布式文件架构，具有极强的可靠性。

2) 高性能。

HBase 是面向实时查询的分布式数据库，能够非常高效地查询和写入数据，实现高并发和实时处理数据。

3) 弹性可扩展。

HBase 建立在 Hadoop 的 HDFS 之上，通过线性方式从下到上灵活地增删节点来进行扩展，并且被众多企业广泛地使用在缓存服务器方面。

4) 面向列的操作。

HBase 面向列来进行存储和查询的，包括行键（Row Key）、列族（Column Family）、列修饰符（Column Qualifier）、数据（Value）、时间戳（TimeStamp）和类型（Type），如图 6-12 所示。

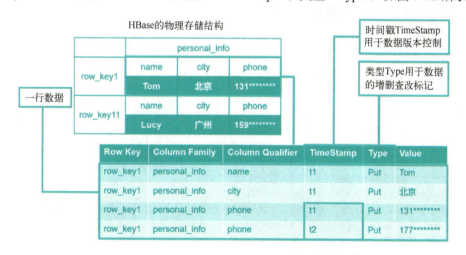

图 6-12 HBase 物理存储结构

(3) HBase 的组织架构（如图 6-13 所示）

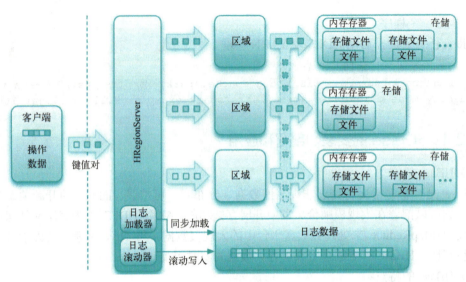

图 6-13　HBase 的组织架构

1）Zookeeper：主要用于为 HBase 提供失败处理机制，选取 HMaster，实时监控 HRegionServer 并通知 HMaster 上下线实时信息，保存 HBase 的主题、表格和列族信息。

2）HMaster：用于管理 HRegionServer 对 Region 的分配和负载均衡。HMaster 中拥有创建 HRegion 的元数据。

3）HRegionServer：用于维护 Master 创建的 HRegion。一个 HRegion 表示一定范围的行键。

4）Store：一个或多个 Store 表示一个 region，一个列族 Column Family 由一个 Store 表示。如果多个 Store 合在一起，HBase 将考虑会切分 region。

5）MemStore：用于保存修改的 HBase 键值 KeyValue 数据，并存储在内存中。

6）StoreFile：表示 MemStore 写入文件后的状态，是 HFile 的轻量级封装。

7）HFile：Hadoop 的二进制格式文件，表示 HBase 键值 KeyValue 数据的存储格式。

8）HLog：用于灾难恢复的日志数据。

（4）HBase 部署准备

1）Zookeeper 正常部署。

首先保证 Zookeeper 集群的正常部署，并将其启动。

```
[Hadoop@Hadoop102 ~]$ zk.sh start
-------启动 Hadoop102 zookeeper-------
ZooKeeper JMX enabled by default
Using config: /opt/module/zookeeper-3.4.10/bin/../conf/zoo.cfg
Starting zookeeper ... already running as process 19097.
-------启动 Hadoop103 zookeeper-------
ZooKeeper JMX enabled by default
Using config: /opt/module/zookeeper-3.4.10/bin/../conf/zoo.cfg
Starting zookeeper ... already running as process 93858.
-------启动 Hadoop104 zookeeper-------
ZooKeeper JMX enabled by default
Using config: /opt/module/zookeeper-3.4.10/bin/../conf/zoo.cfg
Starting zookeeper ... already running as process 97253.
[Hadoop@Hadoop102 kafka]$ jps
119104 NameNode
```

```
119267 DataNode
52774 Jps
19097 QuorumPeerMain
31215 Kafka
```

2）Hadoop 正常部署。

Hadoop 集群的正常部署与启动。

```
[Hadoop@Hadoop102 ~]$ sbin/start-dfs.sh
[Hadoop@Hadoop103 ~]$ sbin/start-yarn.sh
```

（5）HBase 部署

1）HBase 的解压。

使用 tar -zxvf 命令解压 HBase 到指定目录/opt/module/。

```
[Hadoop@Hadoop102 software]$ tar -zxvf hbase-1.3.1-bin.tar.gz -C /opt/module/
```

使用 mv 命令将目录 hbase-1.3.1 的名字改名为 hbase。

```
[Hadoop@Hadoop102 module]$ mv hbase-1.3.1/ hbase
```

2）HBase 的配置。

修改 HBase 对应的配置文件。

hbase-env.sh 的修改内容如下。

```
# The java implementation to use. Java 1.7+ required.
export JAVA_HOME=/opt/module/jdk1.8.0_144
#让 HBase 使用指定的 Zookeeper,如果不改为 false,HBase 会修改 Zookeeper 中的 myid,
导致下次 Zookeeper 启动失败
export HBASE_MANAGES_ZK=false

#注释它们,不注释也没关系,只不过会出现警告
# Configure PermSize. Only needed in JDK7. You can safely remove it for JDK8+
#export HBASE_MASTER_OPTS="$HBASE_MASTER_OPTS -XX:PermSize=128m -XX:Max-
PermSize=128m"
#export HBASE_REGIONSERVER_OPTS="$HBASE_REGIONSERVER_OPTS -XX:PermSize=
128m -XX:MaxPermSize=128m"
```

hbase-site.xml 的修改内容如下。

```
<configuration>
<property>
        <name>hbase.rootdir</name>
        <value>hdfs://Hadoop102:9000/HBase</value>
</property>
<property>
        <name>hbase.cluster.distributed</name>
        <value>true</value>
</property>
<!-- 0.98 后的变动,之前的版本没有.port,默认端口为 60000 -->
<property>
        <name>hbase.master.port</name>
        <value>16000</value>
</property>
<property>
```

```
        <name>hbase.zookeeper.quorum</name>
        <value>Hadoop102,Hadoop103,Hadoop104</value>
</property>
<property>
        <name>hbase.zookeeper.property.dataDir</name>
        <value>/opt/module/zookeepe-3.4.10/zkData</value>
</property>

</configuration>
```

regionservers 文件的修改内容如下。

```
Hadoop102
Hadoop103
Hadoop104
```

将 Hadoop 配置文件软连接到 HBase。

```
[Hadoop@Hadoop102 module]$ ln -s /opt/module/Hadoop-2.7.2/etc/Hadoop/core-site.xml /opt/module/hbase/conf/core-site.xml
[Hadoop@Hadoop102 module]$ ln -s /opt/module/Hadoop-2.7.2/etc/Hadoop/hdfs-site.xml /opt/module/hbase/conf/hdfs-site.xml
```

配置环境变量。

```
[Hadoop@Hadoop102 hbase]$ sudo vim /etc/profile.d/env.sh

##HBASE_HOME
export HBASE_HOME=/opt/module/hbase
export PATH=$PATH:$HBASE_HOME/bin

[Hadoop@Hadoop102 hbase]$ source /etc/profile.d/env.sh
```

HBase 远程发送到其他集群。

```
[Hadoop@Hadoop102 module]$ xsync /opt/module/hbase/
```

分发环境变量文件。

```
[root@Hadoop102 hbase]$ sudo xsync /etc/profile.d/env.sh
```

(6) HBase 服务的启动

1) 启动方式 1。

```
[Hadoop@Hadoop102 hbase]$ bin/hbase-daemon.sh start master
[Hadoop@Hadoop102 hbase]$ bin/hbase-daemon.sh start regionserver
```

2) 启动方式 2。

```
[Hadoop@Hadoop102 hbase]$ bin/start-hbase.sh
```

对应的停止服务。

```
[Hadoop@Hadoop102 hbase]$ bin/stop-hbase.sh
```

(7) 查看 HBase 页面

启动成功后，可以通过"host:port"的方式来访问 HBase 管理页面，即http://Hadoop102:16010/，如图 6-14 所示。

图 6-14　查看 HBase 页面

（8）HBase Shell 操作

1）基本操作。

进入 HBase 客户端命令行。

```
[Hadoop@Hadoop102 hbase]$ bin/hbase shell
```

查看帮助命令。如图 6-15 所示。

```
hbase(main):001:0> help
```

图 6-15　查看帮助命令

查看当前数据库中有哪些表。

```
hbase(main):002:0> list
```

2）表的操作。

创建表。

```
hbase(main):003:0> create 'student','info'
```

插入数据到表中。

```
hbase(main):004:0> put 'student','1001','info:sex','male'
hbase(main):011:0> put 'student','1001','info:age','18'
hbase(main):012:0> put 'student','1002','info:name','Janna'
hbase(main):013:0> put 'student','1002','info:sex','female'
hbase(main):015:0> put 'student','1002','info:age','20'
```

扫描查看表数据。

```
hbase(main):018:0> scan 'student'
hbase(main):024:1> scan 'student',{STARTROW => '1001',STOPROW => '1001'}
```

```
hbase(main):002:0> scan 'student',{STARTROW => '1001'}
```
查看表结构。
```
hbase(main):003:0> describe 'student'
```
更新指定字段的数据。
```
hbase(main):004:0> put 'student','1001','info:name','Nick'
hbase(main):005:0> put 'student','1001','info:age','30'
```
查看"指定行"或"指定列族:列"的数据。
```
hbase(main):001:0> get 'student','1001'
hbase(main):002:0> get 'student','1001','info:name'
```
统计表数据行数。
```
hbase(main):011:0> count 'student'
```
变更表信息。

将 info 列族中的数据存放 3 个版本。
```
hbase(main):029:0> alter 'student',{NAME=>'info',VERSIONS=>3}
hbase(main):032:0> get 'student','1001',{COLUMN=>'info:name',VERSIONS=>3}
```
删除数据。

删除某 rowkey 的全部数据。
```
hbase(main):012:0> deleteall 'student','1001'
```
删除某 rowkey 的某一列数据。
```
hbase(main):013:0> delete 'student','1002','info:sex'
```
清空表数据。
```
hbase(main):014:0> truncate 'student'
```
删除表。

首先需要让该表为 disable 状态。
```
hbase(main):015:0> disable 'student'
```
然后才能使用 drop 命令删除这个表。
```
hbase(main):016:0> drop 'student'
```

 提示：如果直接使用 drop 命令删除表，会报 "ERROR: Table student is enabled. Disable it first." 错误。

6.2.5 Kafka 的安装、部署和应用

（1）Kafka 简介

Apache Kafka 是一个开放源代码的分布式事件流平台，成千上万的公司使用它来实现高性能数据管道、流分析、数据集成等关键任务应用程序。

（2）Kafka 的特点

1）Kafka 能够类似于消息队列一样发布和订阅消息流。

2）Kafka 能够通过文件方式实现消息流的容错处理。

3）Kafka 能够在消息流实施过程中对消息流进行操作处理。

因此，Kafka 不仅能够在系统或应用程序之间构建可靠的数据流传输管道，还能通过构建实时数据流处理程序来转换和应对数据流。

（3）Kafka 的组织架构（如图 6-16 所示）

1）生产者 Producer：用于向 Kafka 集群以 Topic 的方式发送消息。

2）主题 Topic：一个 Topic 类似于一个消息流的名字。

3）消费者 Consumer：用于不断地从 Kafka 集群接收并处理消息流。

（4）Kafka 安装部署

1）分别对主机 Hadoop102、Hadoop103 和 Hadoop104 实施集群部署规划，如表 6-2 所示。

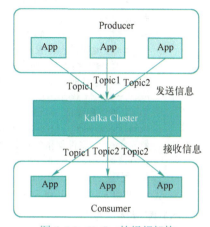

图 6-16　Kafka 的组织架构

表 6-2　集群部署规划

主机名 组件名	Hadoop102	Hadoop103	Hadoop104
Zookeeper	Zookeeper	Zookeeper	Zookeeper
Kafka	Kafka	Kafka	Kafka

2）jar 包下载地址为：http://kafka.apache.org/downloads.html。

（5）集群部署

1）解压安装包。

[Hadoop@Hadoop102 software]$ tar -zxvf kafka_2.11-0.11.0.0.tgz -C /opt/module/

2）修改解压后的文件名称。

[Hadoop@Hadoop102 module]$ mv kafka_2.11-0.11.0.0/ kafka

3）在/opt/module/kafka 目录下创建 logs 文件夹。

[Hadoop@Hadoop102 kafka]$ mkdir logs

4）修改配置文件。

[Hadoop@Hadoop102 config]$ vim server.properties

修改以下内容。

```
#broker 的全局唯一编号，不能重复
broker.id=0
#设置为 true，允许启动删除 topic 的功能
delete.topic.enable=true
#处理网络请求的线程数量
num.network.threads=3
#处理磁盘 IO 的线程数量
num.io.threads=8
#发送套接字的缓冲区大小
socket.send.buffer.bytes=102400
```

```
#接收套接字的缓冲区大小
socket.receive.buffer.bytes=102400
#请求套接字的缓冲区大小
socket.request.max.bytes=104857600
#kafka 运行日志存放的路径
log.dirs=/opt/module/kafka/logs
#topic 在当前 broker 上的分区个数
num.partitions=1
#用来恢复和清理 data 下数据的线程数量
num.recovery.threads.per.data.dir=1
#segment 文件保留的最长时间，超时将被删除
log.retention.hours=168
#配置连接 Zookeeper 的集群地址
zookeeper.connect=Hadoop102:2181,Hadoop103:2181,Hadoop104:2181
```

5）配置环境变量。

```
[Hadoop@Hadoop102 kafka]$ sudo vim /etc/profile.d/env.sh

##KAFKA_HOME
export KAFKA_HOME=/opt/module/kafka
export PATH=$PATH:$KAFKA_HOME/bin

[Hadoop@Hadoop102 kafka]$ source /etc/profile.d/env.sh
```

6）分发安装包。

```
[Hadoop@Hadoop102 module]$ xsync kafka/
```

7）分别在 Hadoop103 和 Hadoop104 上修改配置文件/opt/module/kafka/config/server.properties 中的 broker.id=1、broker.id=2（注：broker.id 不得重复）。

```
[Hadoop@Hadoop103 ~]$ vim /opt/module/kafka/config/server.properties
broker.id=1
[Hadoop@Hadoop104 ~]$ vim /opt/module/kafka/config/server.properties
broker.id=2
```

8）启动集群。

依次在 Hadoop102、Hadoop103、Hadoop104 节点上启动 Kafka。

```
[Hadoop@Hadoop102 kafka]$ bin/kafka-server-start.sh -daemon config/server.properties
[Hadoop@Hadoop103 kafka]$ bin/kafka-server-start.sh -daemon config/server.properties
[Hadoop@Hadoop104 kafka]$ bin/kafka-server-start.sh -daemon config/server.properties
```

9）关闭集群。

```
[Hadoop@Hadoop102 kafka]$ bin/kafka-server-stop.sh stop
[Hadoop@Hadoop103 kafka]$ bin/kafka-server-stop.sh stop
[Hadoop@Hadoop104 kafka]$ bin/kafka-server-stop.sh stop
```

10）日志查看。

如果集群启动不能通过 jps 查看到进程，可以通过日志来查看错误信息，但是 Kafka 的日志命名不是像 Hive 一样的 hive.log 命令方式。server.log 为 Kafka 的启动日志信息，在前面的配

项目 6　基于 Hadoop 的数据分析

置中也有修改，如图 6-17 所示。

图 6-17　Kafka 日志查看

（6）Kafka 消费消息

1）查看当前服务器中的所有 Topic。

```
[Hadoop@Hadoop102 kafka]$ bin/kafka-topics.sh --zookeeper Hadoop102:2181
--list
```

2）创建 Topic。

```
[Hadoop@Hadoop102 kafka]$ bin/kafka-topics.sh --zookeeper Hadoop102:2181
--create --topic first --partitions 3 --replication-factor 1
    Created topic "first".
```

选项说明如下。
- topic：定义 Topic 名。
- partitions：定义分区数。
- replication-factor：定义副本数。

Topic 是逻辑上的概念，而 Partition 是物理上的概念，每个 Partition 对应一个 log 文件，该 log 文件中存储的就是 Producer 生产的数据。

3）删除 Topic。

```
[Hadoop@Hadoop102 kafka]$ bin/kafka-topics.sh --delete --topic first --zookeeper Hadoop102:2181
    Topic first is marked for deletion.
    Note: This will have no impact if delete.topic.enable is not set to true.
```

需要在 server.properties 中设置 delete.topic.enable=true，否则只是标记删除。

4）发送消息。

开启发送消息后，不要发送消息。从 Hadoop103 开启消费信息，从 Hadoop102 发送消息就可在 Hadoop03 看到发送的具体内容。

```
[Hadoop@Hadoop102 kafka]$ bin/kafka-console-producer.sh --broker-list Hadoop102:9092 --topic first
>kafka
>abes abes
```

5）消费消息。

```
[Hadoop@Hadoop103 kafka]$ bin/kafka-console-consumer.sh --zookeeper Hadoop102:2181 --topic first
    Using the ConsoleConsumer with old consumer is deprecated and will be removed
```

```
in a future major release. Consider using the new consumer by passing [bootstrap-
server] instead of [zookeeper].
    kafka
    abes abes
```

加入 --from-beginning 会把主题中所有的数据都读取出来。

```
    [Hadoop@Hadoop104 kafka]$ bin/kafka-console-consumer.sh --zookeeper Hadoop102:
2181 --topic first --from-beginning
    Using the ConsoleConsumer with old consumer is deprecated and will be removed
in a future major release. Consider using the new consumer by passing [bootstrap-
server] instead of [zookeeper].
    kafka
    abes abes
```

6.2.6 Flume 的安装、部署和应用

（1）Flume 简介

Flume 是一种分布式、可靠且可用的服务，用于有效地收集、聚合和移动大量日志数据。它具有基于流式数据的处理体系结构。它具有可调整、可靠性机制，以及许多故障转移和恢复机制，具有强大的功能和容错能力。它使用一个简单的可扩展数据模型，允许在线分析应用程序。

（2）Flume 的特点

1）可靠性。

基于 Hadoop 的 HDFS 分布式文件系统集群，Flume 传送的日志数据在遇到单节点失败时，能够将数据传送到其他节点，避免了数据丢失。

2）可扩展性。

Flume 采用三层架构：Agent、Collector 和 Storage。每一层均可以水平扩展，并由一个或多个 Master 统一监控、维护和管理。

3）可管理性。

Agent 和 Collector 由 Master 统一管理，多个 Master 又由 ZooKeeper 统一管理，在 Master 上可以通过 Web 和 shell Script Command 两种形式对数据流进行管理，并可以对各个数据源配置和动态加载。

（3）Flume 的组织架构（如图 6-18 所示）

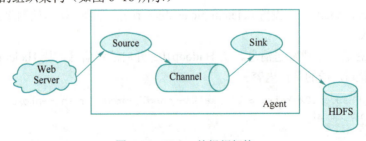

图 6-18　Flume 的组织架构

1）Source：数据的来源和方式。

2）Channel：数据的缓冲池。

3）Sink：定义了数据输出的方式和目的地。

Flume 的关键流程是首先通过 Source 获取数据源的数据，然后将数据缓存在 Channel 中以

保证数据在传输过程中不丢失，最后通过 Sink 将数据发送到指定的位置。

（4）下载地址

Flume 官网地址为：http://flume.apache.org/。

文档查看地址为：http://flume.apache.org/FlumeUserGuide.html。

下载地址为：http://archive.apache.org/dist/flume/。

（5）安装部署

```
#将 apache-flume-1.7.0-bin.tar.gz 上传到 Linux 的/opt/software 目录下
#解压 apache-flume-1.7.0-bin.tar.gz 到/opt/module/目录下
[Hadoop@Hadoop102 software]$ tar -zxvf apache-flume-1.7.0-bin.tar.gz -C /opt/module/

#修改 apache-flume-1.7.0-bin 的名称为 flume
[Hadoop@Hadoop102 module]$ mv apache-flume-1.7.0-bin/ flume

#将 flume/conf 下的 flume-env.sh.template 文件修改为 flume-env.sh，并配置 flume-env.sh 文件
[Hadoop@Hadoop102 conf]$ mv flume-env.sh.template flume-env.sh
[Hadoop@Hadoop102 conf]$ vim flume-env.sh
export JAVA_HOME=/opt/module/jdk1.8.0_144

#配置环境变量
[Hadoop@Hadoop102 conf]$ sudo vim /etc/profile.d/env.sh

#追加以下内容
##FLUME_HOME
export FLUME_HOME=/opt/module/flume
export PATH=$PATH:$FLUME_HOME/bin

#刷新环境变量
[Hadoop@Hadoop102 conf]$ source /etc/profile.d/env.sh
```

（6）监控端口数据官方案例

1）案例需求。

使用 Flume 监听一个端口，收集该端口数据，并打印到控制台。

2）需求分析。

通过 netcat 工具向本机 44444 端口发送数据，开启 Flume 监控本机的 44444 端口获取数据，将获取的数据输出到控制台，通过 Flume 的 Source 端获取数据，通过 Sink 端输出，如图 6-19 所示。

图 6-19　监控端口数据流程

3）实现步骤。

安装 netcat 工具。

  [Hadoop@Hadoop102 conf]$ sudo yum install -y nc

判断 44444 端口是否被占用。

  [Hadoop@Hadoop102 conf]$ sudo netstat -tunlp | grep 44444

创建 Flume Agent 配置文件 flume-netcat-logger.conf。

在 flume 目录下创建 job 文件夹并进入 job 文件夹。

```
[Hadoop@Hadoop102 flume]$ mkdir job
#在 job 文件夹下创建 Flume Agent 配置文件 flume-netcat-logger.conf
[Hadoop@Hadoop102 job]$ touch flume-netcat-logger.conf
#在 flume-netcat-logger.conf 文件中添加如下内容
# Name the comonents on this agent
# Source 的名称为 r1
a1.sources = r1
# Sink 的名称为 k1
a1.sinks = k1
# Channels 的名称为 c1
a1.channels = c1

# Describe / configure the source
# a1 的输入源类型为 netcat 端口类型
a1.sources.r1.type = netcat
# a1 监听的主机
a1.sources.r1.bind = 192.168.19.102
# a1 监听的端口
a1.sources.r1.port = 44444

# Describe the sink
# a1 的输出类型为 logger 类型
a1.sinks.k1.type = logger

# Use a channel which buffers events in memory
# a1 的 channel 的类型为 memory 内存型
a1.channels.c1.type = memory
# a1 的 channel 的总容量为 1000 个 event
a1.channels.c1.capacity =1000
# a1 的 channel 收集 100 个 event 再去提交事务
a1.channels.c1.transactionCapacity = 100

# Bind the source and sink to the channel
# 连接 r1 和 c1
a1.sources.r1.channels = c1
# 连接 k1 和 c1
a1.sinks.k1.channel = c1
```

注意：配置文件来源于官方手册http://flume.apache.org/FlumeUserGuide.html。

配置文件解析如图 6-20 所示。

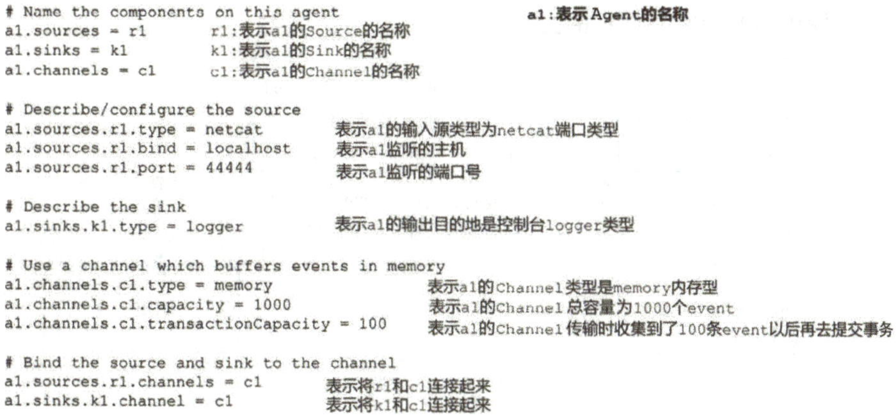

```
# Name the components on this agent                          a1:表示Agent的名称
a1.sources = r1          r1:表示a1的Source的名称
a1.sinks = k1            k1:表示a1的Sink的名称
a1.channels = c1         c1:表示a1的Channel的名称

# Describe/configure the source
a1.sources.r1.type = netcat          表示a1的输入源类型为netcat端口类型
a1.sources.r1.bind = localhost       表示a1监听的主机
a1.sources.r1.port = 44444           表示a1监听的端口号

# Describe the sink
a1.sinks.k1.type = logger            表示a1的输出目的地是控制台logger类型

# Use a channel which buffers events in memory
a1.channels.c1.type = memory                      表示a1的Channel类型是memory内存型
a1.channels.c1.capacity = 1000                    表示a1的Channel总容量为1000个event
a1.channels.c1.transactionCapacity = 100          表示a1的Channel传输时收集到了100条event以后再去提交事务

# Bind the source and sink to the channel
a1.sources.r1.channels = c1          表示将r1和c1连接起来
a1.sinks.k1.channel = c1             表示将k1和c1连接起来
```

图 6-20　flume-netcat-logger.conf 文件配置

先开启 Flume 监听端口。

```
#第一种写法
[Hadoop@Hadoop102 flume]$ bin/flume-ng agent --conf conf/ --name a1
--conf-file job/flume-netcat-logger.conf -Dflume.root.logger=INFO,console

#第二种写法
[Hadoop@Hadoop102 flume]$ bin/flume-ng agent -c conf/ -n a1
-f job/flume-netcat-logger.conf -Dflume.root.logger=INFO,console
```

参数说明如下。

- conf/-c：配置文件的目录。
- name/-n：给 Agent 命名。
- conf-file/-f：Flume 本次启动读取的配置文件。
- -Dflume.root.logger=INFO，console：-D 表示 Flume 运行时动态修改 flume.root.logger 参数属性值，将控制台日志打印级别设置为 INFO 级别。

使用 netcat 工具向本机的 44444 端口发送内容。

```
[Hadoop@Hadoop102 module]$ nc 192.168.19.102 44444
hello
OK
```

用 Flume 监听页面观察接收数据情况，如图 6-21 所示。

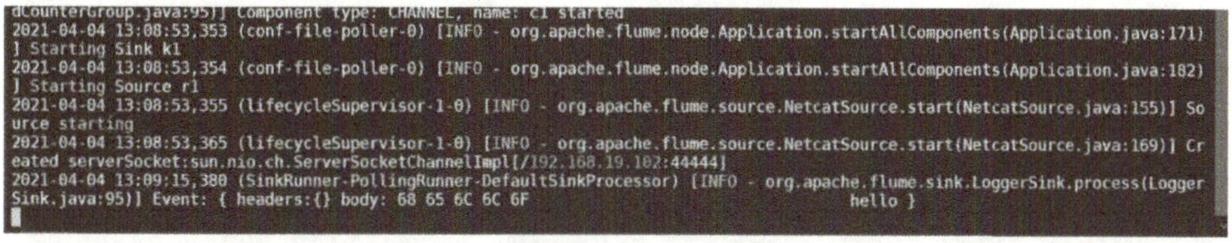

```
dCounterGroup.java:95)] Component type: CHANNEL, name: c1 started
2021-04-04 13:08:53,353 (conf-file-poller-0) [INFO - org.apache.flume.node.Application.startAllComponents(Application.java:171)
] Starting Sink k1
2021-04-04 13:08:53,354 (conf-file-poller-0) [INFO - org.apache.flume.node.Application.startAllComponents(Application.java:182)
] Starting Source r1
2021-04-04 13:08:53,355 (lifecycleSupervisor-1-0) [INFO - org.apache.flume.source.NetcatSource.start(NetcatSource.java:155)] So
urce starting
2021-04-04 13:08:53,365 (lifecycleSupervisor-1-0) [INFO - org.apache.flume.source.NetcatSource.start(NetcatSource.java:169)] Cr
eated serverSocket:sun.nio.ch.ServerSocketChannelImpl[/192.168.19.102:44444]
2021-04-04 13:09:15,380 (SinkRunner-PollingRunner-DefaultSinkProcessor) [INFO - org.apache.flume.sink.LoggerSink.process(Logger
Sink.java:95)] Event: { headers:{} body: 68 65 6C 6C 6F                               hello }
```

图 6-21　Flume 监听页面

（7）实时监控单个追加文件

1）案例需求。

实时监控 Hive 日志，并上传到 HDFS 中。

2）需求分析

创建符合条件的 Flume 配置文件，开启监控，操作 hive，Flume 通过 Source 端获取的数据再通过 Sink 端输出到 HDFS，如图 6-22 所示。

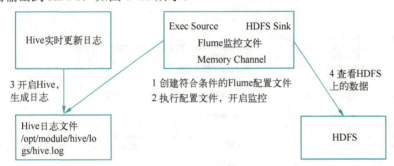

图 6-22　Flume 工作流程

3）实现步骤。

Flume 要想将数据输出到 HDFS，须持有 Hadoop 相关 jar 包。

将以下 jar 包复制到/opt/module/flume/lib 文件夹下。

```
commons-configuration-1.6.jar
Hadoop-auth-2.7.2.jar
Hadoop-common-2.7.2.jar
Hadoop-hdfs-2.7.2.jar
commons-io-2.4.jar
htrace-core-3.1.0-incubating.jar
```

创建 flume-file-hdfs.conf 文件。

```
[Hadoop@Hadoop102 job]$ touch flume-file-hdfs.conf
```

为 flume-file-hdfs.conf 添加如下内容。

```
# Name the components on this agent
a2.sources = r2
a2.sinks = k2
a2.channels = c2

# Describe/configure the source
a2.sources.r2.type = exec
a2.sources.r2.command = tail -F /opt/module/hive/logs/hive.log
a2.sources.r2.shell = /bin/bash -c

# Describe the sink
a2.sinks.k2.type = hdfs
a2.sinks.k2.hdfs.path = hdfs.Hadoop102:9000/flume/%Y%d/%H
# 上传的前缀
a2.sinks.k2.hdfs.filePrefix = logs-
# 是否按照时间滚动文件夹
a2.sinks.k2.hdfs.round = true
# 多少时间单位创建一个新的文件夹
a2.sinks.k2.hdfs.roundValue = 1
# 重新定义时间单位
a2.sinks.k2.hdfs.roundUnit = hour
```

```
# 是否使用本地时间戳
a2.sinks.k2.hdfs.useLocalTimeStamp = true
# 通过批处理方式累积到一定数量的事件（Event）后再一次性地加载到 HDFS
a2.sinks.k2.hdfs.batchSize = 1000
# 设置文件类型，可支持压缩
a2.sinks.k2.hdfs.fileType = DataStream
# 多久生成一个新的文件
a2.sinks.k2.hdfs.rollInterval = 30
# 设置每个文件的滚动大小
a2.sinks.k2.hdfs.rollsize = 134217700
# 文件的滚动与 Event 数量无关
a2.sinks.k2.hdfs.rollCount = 0

# Use a channel which buffers events in memory
a2.channels.c2.type = memory
a2.channels.c2.capacity = 1000
a2.channels.c2.trannsactionCapacity = 100

# Bind the source and sink to the channel
a2.sources.r2.channels = c2
a2.sinks.k2.channel = c2
```

配置文件解析如图 6-23 所示。

图 6-23 flume-netcat-logger.conf 配置文件

运行 Flume。

```
[Hadoop@Hadoop102 flume]$ bin/flume-ng agent -c conf/ -n a2 -f job/flume-file-hdfs.conf
```

开启 Hadoop 和 Hive 并操作 Hive 产生日志。

```
[Hadoop@Hadoop102 Hadoop-2.7.2]$ start-dfs.sh
[Hadoop@Hadoop102 Hadoop-2.7.2]$ start-yarn.sh
[Hadoop@Hadoop102 hive]$ bin/hive
```

在 HDFS 上查看文件，如图 6-24 所示。

图 6-24　在 HDFS 上查看文件

6.2.7　Sqoop 的安装、部署和应用

（1）Sqoop 简介

Sqoop 主要负责在 Hadoop 的 HDFS 和关系型数据库之间进行数据迁移。Sqoop 的具体实现是由 Mapreduce 程序来完成的。

（2）Sqoop 的特点

1）并行导入/导出。

Sqoop 使用基于 Hadoop 的 YARN 框架，在并行性的基础上也提供容错功能。

2）连接主流 RDBMS 数据库。

Sqoop 可以面向主流的 RDBMS 数据库进行连接。

3）全部加载。

使用 Sqoop 中的命令加载整张表，甚至可以加载数据库中的所有表。

4）直接与 Hive 和 HBase 交互。

可以直接将数据加载到 Hive 和 HBase 中，并可以在 Hive 中压缩数据。

（3）Sqoop 的组织架构（如图 6-25 所示）

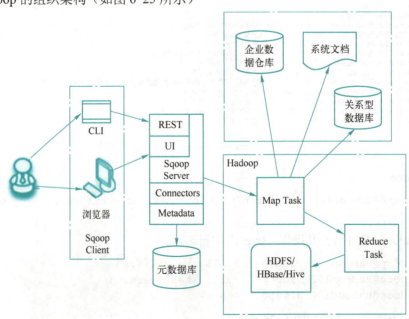

图 6-25　Sqoop 的组织架构

1) Sqoop Server：所有的连接器 Connectors 都安装在 Sqoop Server 上便于统一管理。其中，Connectors 负责数据读写，Metadata 负责管理数据库的元数据。

2) Sqoop Client：通过浏览器或者 CLI 实现客户端的 REST API、Java API、Web UI，以及 CLI 控制台与 Sqoop Server 的交互。

（4）Sqoop 安装部署

1) 下载安装包，如图 6-26 所示。

下载地址为：http://archive.apache.org/dist/sqoop/1.4.6/。

图 6-26 Sqoop 安装包

2) 上传解压。

解压至/opt/module/目录下。

```
[Hadoop@Hadoop102 software]$ tar -zxvf sqoop-1.4.6.bin__Hadoop-2.0.4-alpha.tar.gz -C /opt/module/
```

使用 mv 命令将目录 sqoop-1.4.6.bin__Hadoop-2.0.4-alpha 的名字修改为 sqoop。

```
[Hadoop@Hadoop102 software]$ cd /opt/module/
[Hadoop@Hadoop102 module]$ mv sqoop-1.4.6.bin__Hadoop-2.0.4-alpha/ sqoop
```

下面配置环境变量。

记住 Sqoop 的绝对路径。

```
[Hadoop@Hadoop102 sqoop]$ pwd
/opt/module/sqoop
```

修改环境变量。

```
[Hadoop@Hadoop102 sqoop]$ sudo vim /etc/profile.d/env.sh
```

追加以下内容。

```
##SQOOP_HOME
export SQOOP_HOME=/opt/module/sqoop
export $PATH=$PATH:$SQOOP_HOME/bin
```

刷新环境变量。

```
[Hadoop@Hadoop102 sqoop]$ source /etc/profile.d/env.sh
```

分发环境变量。

```
[Hadoop@Hadoop102 sqoop]$ su
[root@Hadoop102 sqoop]# xsync /etc/profile.d/env.sh
```

3）修改配置。

将 sqoop-env-template.sh 重命名为 sqoop-env.sh。

```
[Hadoop@Hadoop102 conf]$ mv sqoop-env-template.sh sqoop-env.sh
```

修改 sqoop-env.sh。

设置 sqoop 的环境变量。

```
# Set Hadoop-specific environment variables here.

#Set path to where bin/Hadoop is available
export Hadoop_COMMON_HOME=/opt/module/Hadoop-2.7.2

#Set path to where Hadoop-*-core.jar is available
export Hadoop_MAPRED_HOME=/opt/module/Hadoop-2.7.2

#set the path to where bin/hbase is available
export HBASE_HOME=/opt/module/hbase

#Set the path to where bin/hive is available
export HIVE_HOME=/opt/module/hive

#Set the path for where zookeper config dir is
export ZOOCFGDIR=/opt/module/zookeeper-3.4.10
```

4）复制 JDBC 驱动。

```
[Hadoop@Hadoop102 lib]$ pwd
/opt/module/sqoop/lib

[Hadoop@Hadoop102 lib]$ cp /opt/software/mysql-libs/mysql-connector-java-5.1.27/mysql-connector-java-5.1.27-bin.jar ./
```

5）验证 Sqoop。

验证 Sqoop 是否配置正确。

```
[Hadoop@Hadoop102 sqoop]$ bin/sqoop help
```

出现 Warning 警告和帮助信息。

```
    Warning: /opt/module/sqoop/bin/../../hcatalog does not exist! HCatalog jobs will fail.
    Please set $HCAT_HOME to the root of your HCatalog installation.
    Warning: /opt/module/sqoop/bin/../../accumulo does not exist! Accumulo imports will fail.
    Please set $ACCUMULO_HOME to the root of your Accumulo installation.
    21/04/06 14:50:45 INFO sqoop.Sqoop: Running Sqoop version: 1.4.6
    usage: sqoop COMMAND [ARGS]

    Available commands:
      codegen            Generate code to interact with database records
      create-hive-table  Import a table definition into Hive
```

```
eval               Evaluate a SQL statement and display the results
export             Export an HDFS directory to a database table
help               List available commands
import             Import a table from a database to HDFS
import-all-tables  Import tables from a database to HDFS
import-mainframe   Import datasets from a mainframe server to HDFS
job                Work with saved jobs
list-databases     List available databases on a server
list-tables        List available tables in a database
merge              Merge results of incremental imports
metastore          Run a standalone Sqoop metastore
version            Display version information

See 'sqoop help COMMAND' for information on a specific command.
```

6）测试 Sqoop 是否成功连接数据库。

```
[Hadoop@Hadoop102 sqoop]$ bin/sqoop list-databases --connect jdbc:mysql:
//Hadoop102:3306 --username root --password 123456
```

出现如下信息。

```
Warning: /opt/module/sqoop/bin/../../hcatalog does not exist! HCatalog jobs will fail.
Please set $HCAT_HOME to the root of your HCatalog installation.
Warning: /opt/module/sqoop/bin/../../accumulo does not exist! Accumulo imports will fail.
Please set $ACCUMULO_HOME to the root of your Accumulo installation.
21/04/06 14:56:00 INFO sqoop.Sqoop: Running Sqoop version: 1.4.6
21/04/06 14:56:00 WARN tool.BaseSqoopTool: Setting your password on the command-line is insecure. Consider using -P instead.
21/04/06 14:56:00 INFO manager.MySQLManager: Preparing to use a MySQL streaming resultset.
information_schema
metastore
mysql
performance_schema
test
```

可以看到数据库说明连接成功了。

（5）Sqoop 的数据导入使用案例

Sqoop 中，"导入"指从非大数据集群（RDBMS）向大数据集群（HDFS、Hive、HBase）中传输数据。

1）将数据从 RDBMS 导入到 HDFS。

确定 MySQL 服务开启正常。

在 MySQL 中新建一张表并插入一些数据。

```
$ mysql -uroot -p000000
mysql> create database company;
mysql> create table company.staff(id int(4) primary key not null auto_increment, name varchar(255), sex varchar(255));
mysql> insert into company.staff(name, sex) values('Tom', 'Male');
```

```
mysql> insert into company.staff(name, sex) values('lila', 'FeMale');
```

2）全部导入。

```
$ bin/sqoop import \
--connect jdbc:mysql://Hadoop102:3306/company \
--username root \
--password 000000 \
--table staff \
--target-dir /user/company \
--delete-target-dir \
--num-mappers 1 \
--fields-terminated-by "\t"
```

3）查询导入。

```
$ bin/sqoop import \
--connect jdbc:mysql://Hadoop102:3306/company \
--username root \
--password 000000 \
--target-dir /user/company \
--delete-target-dir \
--num-mappers 1 \
--fields-terminated-by "\t" \
--query 'select name,sex from staff where id <=1 and $CONDITIONS;'
```

4）导入指定列。

```
$ bin/sqoop import \
--connect jdbc:mysql://Hadoop102:3306/company \
--username root \
--password 000000 \
--target-dir /user/company \
--delete-target-dir \
--num-mappers 1 \
--fields-terminated-by "\t" \
--columns id,sex \
--table staff
```

5）使用 sqoop 关键字筛选查询导入数据。

```
$ bin/sqoop import \
--connect jdbc:mysql://Hadoop102:3306/company \
--username root \
--password 000000 \
--target-dir /user/company \
--delete-target-dir \
--num-mappers 1 \
--fields-terminated-by "\t" \
--table staff \
--where "id=1"
```

6）使用 sqoop 的 import 命令将数据从 RDBMS 导入到 Hive。

```
$ bin/sqoop import \
```

```
--connect jdbc:mysql://Hadoop102:3306/company \
--username root \
--password 000000 \
--table staff \
--num-mappers 1 \
--hive-import \
--fields-terminated-by "\t" \
--hive-overwrite \
--hive-table staff_hive
```

7）使用 sqoop 的 import 命令将数据从 RDBMS 导入到 HBase。

```
$ bin/sqoop import \
--connect jdbc:mysql://Hadoop102:3306/company \
--username root \
--password 000000 \
--table staff \
--num-mappers 1 \
--hive-import \
--fields-terminated-by "\t" \
--hive-overwrite \
--hive-table staff_hive
```

提示：需手动创建 HBase 表。

```
hbase> create 'hbase_company,'info'
```

8）在 HBase 中 scan 表得到如下内容。

```
hbase> scan 'hbase_company'
```

（6）Sqoop 的数据导出使用案例

在 Sqoop 中，"导出"指从大数据集群（HDFS、Hive、HBase）向非大数据集群（RDBMS）中传输数据。

将数据从 HIVE/HDFS 导入到 RDBMS。

```
$ bin/sqoop export \
--connect jdbc:mysql://Hadoop102:3306/company \
--username root \
--password 000000 \
--table staff \
--num-mappers 1 \
--export-dir /user/hive/warehouse/staff_hive \
--input-fields-terminated-by "\t"
```

提示：MySQL 中如果表不存在，不会自动创建。

6.2.8 Zookeeper 的安装、部署和应用

（1）Zookeeper 简介

Zookeeper 是一套能够在大规模分布式系统当中实现系统内部各应用程序的调节和服务的工作系统。

(2) Zookeeper 的特点。

1) 顺序一致性。

一个服务器发起的消息顺序，所有服务器按同样顺序执行。

2) 视图一致性。

客户端在一台服务器看到的数据视图与其他服务器的视图是一致的。

3) 可靠性。

简单、稳定、高效的操作方式让消息被一台服务器接收后也能被所有其他服务器接收。

4) 实时性。

Zookeeper 具有一定的数据实时同步功能，保障各个服务器的数据能够及时传送。

5) 原子性。

处理的事务只有成功或者失败两种情况，没有第三种情况。

(3) Zookeeper 的组织架构（如图 6-27 所示）

1) Leader：所有 Zookeeper 服务器中只有一个 Leader 被选举出来，作为整个 ZooKeeper 集群的主节点，其他节点都是 Follower 或 Observer。Leader 是所有应用程序事务请求的最高协调和决定者，统一管理集群事物的执行顺序，保证整个集群内部消息处理的先进先出。

2) Observer：主要应用于需要处理更多负载或者跨机房的应用场景，用于提升系统可扩展性，提升读取速度，但 Observer 不参与选举投票。

3) Follower：接收 Client 的请求，返回响应给 Client，并参与 Leader 的选举投票。

4) Client:事务请求发送者。

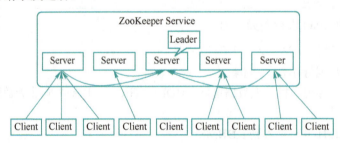

图 6-27 Zookeeper 的组织架构

(4) 集群规划

在 Hadoop102、Hadoop103 和 Hadoop104 三个节点上部署 Zookeeper。

(5) 解压安装

解压 Zookeeper 安装包到/opt/module/目录下。

```
[Hadoop@Hadoop102 software]$ tar -zxvf zookeeper-3.4.10.tar.gz -C /opt/module/
```

同步/opt/module/zookeeper-3.4.10 目录内容到 Hadoop103、Hadoop104。

```
[Hadoop@Hadoop102 module]$ xsync zookeeper-3.4.10/
```

(6) 配置服务器编号

在/opt/module/zookeeper-3.4.10/目录下创建 zkData。

```
[Hadoop@Hadoop102 zookeeper-3.4.10]$ mkdir -p zkData
```

在/opt/module/zookeeper-3.4.10/zkData 目录下创建一个名为 myid 的文件。

```
[Hadoop@Hadoop102 zkData]$ touch myid
```

编辑 myid 文件。

```
[Hadoop@Hadoop102 zkData]$ vim myid
```

在文件中添加与 server 对应的编号 2。

复制配置好的 Zookeeper 到其他机器上。

```
[Hadoop@Hadoop102 zookeeper-3.4.10]$ xsync zkData/
#分别在 Hadoop102、Hadoop103 上修改 myid 文件中的内容为 3、4
```

（7）配置 zoo.cfg 文件

重命名 /opt/module/zookeeper-3.4.10/conf 目录下的 zoo_sample.cfg 为 zoo.cfg。

```
[Hadoop@Hadoop102 conf]$ mv zoo_sample.cfg zoo.cfg
```

打开 zoo.cfg 文件。

```
[Hadoop@Hadoop102 conf]$ vim zoo.cfg
#修改数据存储路径配置
dataDir=/opt/module/zookeeper-3.4.10/zkData
#增加如下配置
#######################cluster#########################
server.2=Hadoop102:2888:3888
server.3=Hadoop103:2888:3888
server.4=Hadoop104:2888:3888
```

同步 zoo.cfg 配置文件。

```
[Hadoop@Hadoop102 conf]$ xsync zoo.cfg
```

（8）集群操作

1）分别启动 Zookeeper。

```
[Hadoop@Hadoop102 zookeeper-3.4.10]$ bin/zkServer.sh start
ZooKeeper JMX enabled by default
Using config: /opt/module/zookeeper-3.4.10/bin/../conf/zoo.cfg
Starting zookeeper ... STARTED

[Hadoop@Hadoop103 zookeeper-3.4.10]$ bin/zkServer.sh start
ZooKeeper JMX enabled by default
Using config: /opt/module/zookeeper-3.4.10/bin/../conf/zoo.cfg
Starting zookeeper ... STARTED

[Hadoop@Hadoop104 zookeeper-3.4.10]$ bin/zkServer.sh stop
ZooKeeper JMX enabled by default
Using config: /opt/module/zookeeper-3.4.10/bin/../conf/zoo.cfg
Stopping zookeeper ... STOPPED
```

2）查看状态。

```
[Hadoop@Hadoop102 zookeeper-3.4.10]$ bin/zkServer.sh status
ZooKeeper JMX enabled by default
Using config: /opt/module/zookeeper-3.4.10/bin/../conf/zoo.cfg
Mode: follower
```

```
[Hadoop@Hadoop103 zookeeper-3.4.10]$ bin/zkServer.sh status
ZooKeeper JMX enabled by default
Using config: /opt/module/zookeeper-3.4.10/bin/../conf/zoo.cfg
Mode: leader

[Hadoop@Hadoop104 zookeeper-3.4.10]$ bin/zkServer.sh status
ZooKeeper JMX enabled by default
Using config: /opt/module/zookeeper-3.4.10/bin/../conf/zoo.cfg
Mode: follower
```

3）关闭 Zookeeper。

```
[Hadoop@Hadoop102 zookeeper-3.4.10]$ bin/zkServer.sh stop
ZooKeeper JMX enabled by default
Using config: /opt/module/zookeeper-3.4.10/bin/../conf/zoo.cfg
Stopping zookeeper ... STOPPED

[Hadoop@Hadoop103 zookeeper-3.4.10]$ bin/zkServer.sh stop
ZooKeeper JMX enabled by default
Using config: /opt/module/zookeeper-3.4.10/bin/../conf/zoo.cfg
Stopping zookeeper ... STOPPED

[Hadoop@Hadoop104 zookeeper-3.4.10]$ bin/zkServer.sh stop
ZooKeeper JMX enabled by default
Using config: /opt/module/zookeeper-3.4.10/bin/../conf/zoo.cfg
Stopping zookeeper ... STOPPED
```

【任务实施】

任务 1　使用 Hadoop 及其组件 Hive 实现数据分析

（1）计算每个店铺的当月销售额和累积到当月的总销售额。

1）数据准备，如图 6-28 所示。三列数据分别是店铺、月份和金额。

```
[hadoop@hadoop102 datas]$ pwd
/home/hadoop/datas
[hadoop@hadoop102 datas]$ cat shop
a,01,150
a,01,200
b,01,1000
b,01,800
c,01,250
c,01,220
b,01,6000
a,02,2000
a,02,3000
b,02,1000
[hadoop@hadoop102 datas]$ 
```

图 6-28　数据准备

2）创建表格并加载数据，如图 6-29 和图 6-30 所示。

```
hive (default)> create table shop(id string,month string,money int) row
```

format delimited fields terminated by ',';

```
hive (default)> create table shop(id string,month string,money int) row format delimited fields terminated by ',';
OK
Time taken: 0.307 seconds
hive (default)>
```

图 6-29　创建 shop 表

　　hive (default)> load data local inpath '/home/Hadoop/datas/shop' into table shop;

```
hive (default)> load data local inpath '/home/hadoop/datas/shop' into table shop;
Loading data to table default.shop
Table default.shop stats: [numFiles=1, totalSize=95]
OK
Time taken: 0.731 seconds
hive (default)>
```

图 6-30　加载数据到 shop 表

3）先求每个店铺每个月的销售额，如图 6-31 所示。

　　hive (default)> select id,month,sum(money) from shop group by id,month;

```
hive (default)> select id,month,sum(money) from shop group by id,month;
Query ID = hadoop_20210525185441_c7070966-5452-47b7-bfc4-ef9713fa3076
Total jobs = 1
Launching Job 1 out of 1
Number of reduce tasks not specified. Estimated from input data size: 1
In order to change the average load for a reducer (in bytes):
  set hive.exec.reducers.bytes.per.reducer=<number>
In order to limit the maximum number of reducers:
  set hive.exec.reducers.max=<number>
In order to set a constant number of reducers:
  set mapreduce.job.reduces=<number>
Job running in-process (local Hadoop)
2021-05-25 18:54:42,930 Stage-1 map = 100%,  reduce = 100%
Ended Job = job_local599692105_0001
MapReduce Jobs Launched:
Stage-Stage-1:  HDFS Read: 380 HDFS Write: 190 SUCCESS
Total MapReduce CPU Time Spent: 0 msec
OK
id      month   _c2
a       01      350
a       02      5000
b       01      7800
b       02      1000
c       01      470
Time taken: 1.922 seconds, Fetched: 5 row(s)
hive (default)>
```

图 6-31　每个店铺每个月的销售额

4）再求每个店铺的当月销售额和累积到当月的总销售额，如图 6-32 所示。

　　select t1.id,t1.month,t1.m,sum(t1.m) over(partition by id order by month) from (select id,month,sum(money) m from shop group by id,month)t1;

```
MapReduce Jobs Launched:
Stage-Stage-1:  HDFS Read: 570 HDFS Write: 190 SUCCESS
Total MapReduce CPU Time Spent: 0 msec
OK
t1.id   t1.month        t1.m    sum_window_0
a       01      350     350
a       02      5000    5350
b       01      7800    7800
b       02      1000    8800
c       01      470     470
Time taken: 1.462 seconds, Fetched: 5 row(s)
hive (default)>
```

图 6-32　每个店铺的当月销售额和累积到当月的总销售额

（2）求一年中出现最高温度的那一天。

1）数据准备，如图 6-33 所示。

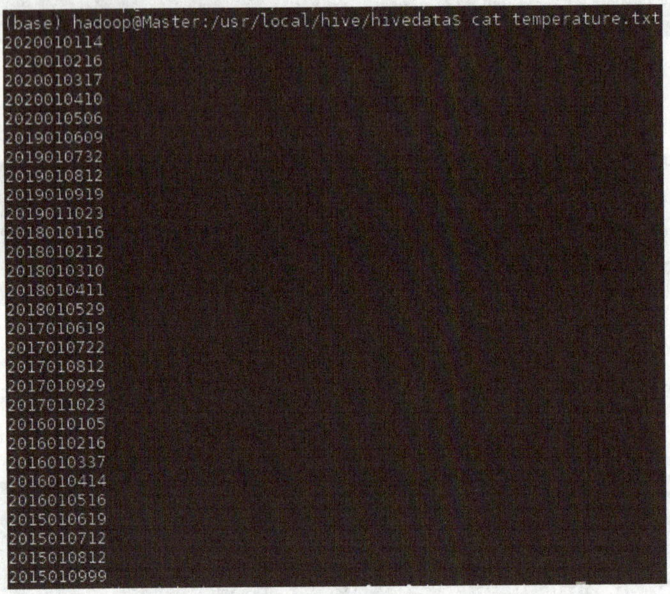

图 6-33　数据准备

数据的前 8 位为日期（2014-01-01），最后两位为温度。

2）创建表格并加载数据，如图 6-34 所示。

```
create table tem(tem string);
load data local inpath '/usr/local/hive/hivedata/temperature.txt' into table tem;
```

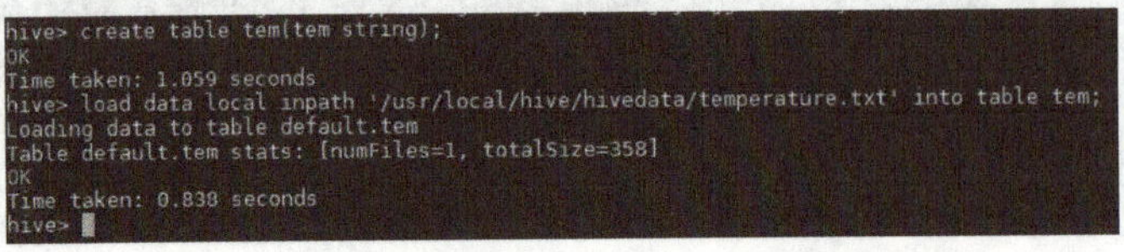

图 6-34　创建表格并加载数据

3）求每年的最高温度，如图 6-35 所示。

```
select substring(t.tem,1,4) as year,max(substring(t.tem,-2)) as max_tem
from tem t
group by substring(t.tem,1,4);
```

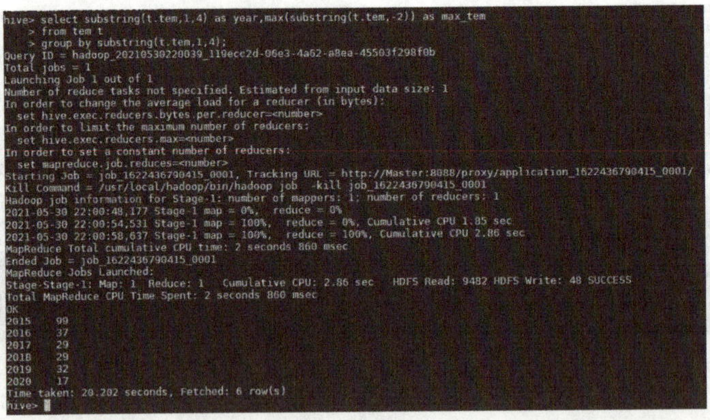

图 6-35　每年的最高温度

4）求一年中出现最高温度的那一天，如图 6-36 所示。

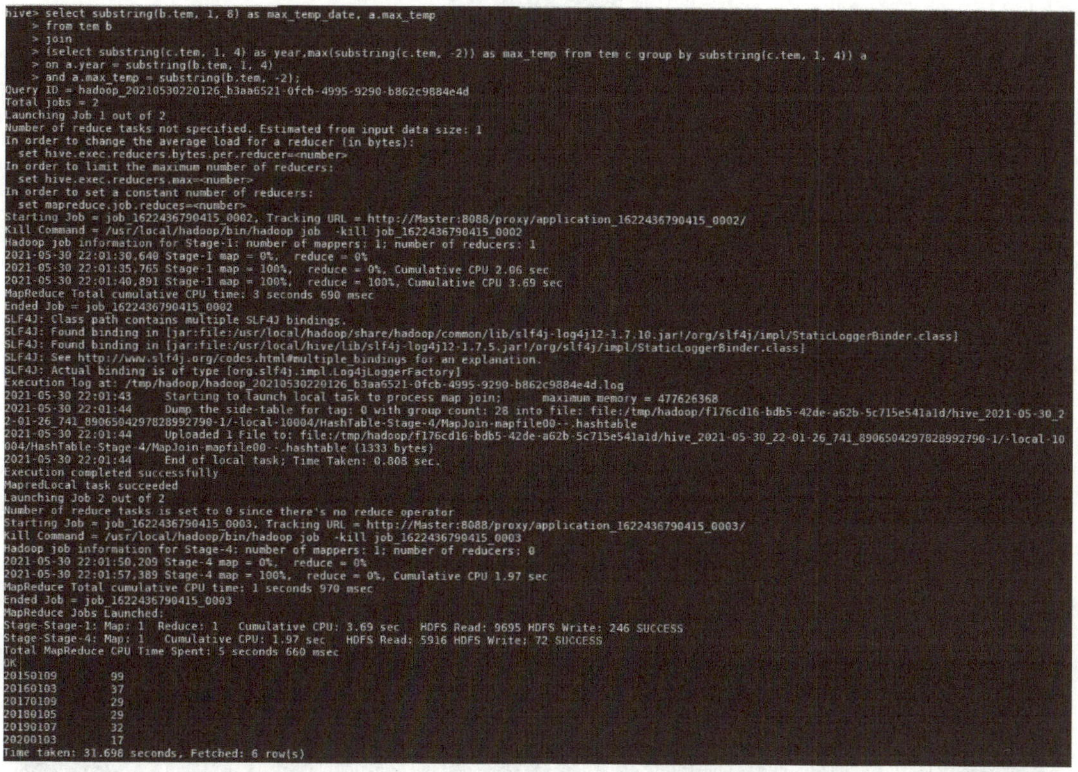

```
hive> select substring(b.tem, 1, 8) as max_temp_date, a.max_temp
    > from tem b
    > join
    > (select substring(c.tem, 1, 4) as year,max(substring(c.tem, -2)) as max_temp from tem c group by substring(c.tem, 1, 4)) a
    > on a.year = substring(b.tem, 1, 4)
    > and a.max_temp = substring(b.tem, -2);
Query ID = hadoop_20210530220126_b3aa6521-0fcb-4995-9290-b862c9884e4d
Total jobs = 2
Launching Job 1 out of 2
Number of reduce tasks not specified. Estimated from input data size: 1
In order to change the average load for a reducer (in bytes):
  set hive.exec.reducers.bytes.per.reducer=<number>
In order to limit the maximum number of reducers:
  set hive.exec.reducers.max=<number>
In order to set a constant number of reducers:
  set mapreduce.job.reduces=<number>
Starting Job = job_1622436790415_0002, Tracking URL = http://Master:8088/proxy/application_1622436790415_0002/
Kill Command = /usr/local/hadoop/bin/hadoop job  -kill job_1622436790415_0002
Hadoop job information for Stage-1: number of mappers: 1; number of reducers: 1
2021-05-30 22:01:30,640 Stage-1 map = 0%,  reduce = 0%
2021-05-30 22:01:35,765 Stage-1 map = 100%,  reduce = 0%, Cumulative CPU 2.06 sec
2021-05-30 22:01:40,891 Stage-1 map = 100%,  reduce = 100%, Cumulative CPU 3.69 sec
MapReduce Total cumulative CPU time: 3 seconds 690 msec
Ended Job = job_1622436790415_0002
SLF4J: Class path contains multiple SLF4J bindings.
SLF4J: Found binding in [jar:file:/usr/local/hadoop/share/hadoop/common/lib/slf4j-log4j12-1.7.10.jar!/org/slf4j/impl/StaticLoggerBinder.class]
SLF4J: Found binding in [jar:file:/usr/local/hive/lib/slf4j-log4j12-1.7.5.jar!/org/slf4j/impl/StaticLoggerBinder.class]
SLF4J: See http://www.slf4j.org/codes.html#multiple_bindings for an explanation.
SLF4J: Actual binding is of type [org.slf4j.impl.Log4jLoggerFactory]
Execution log at: /tmp/hadoop/hadoop_20210530220126_b3aa6521-0fcb-4995-9290-b862c9884e4d.log
2021-05-30 22:01:43     Starting to launch local task to process map join;      maximum memory = 477626368
2021-05-30 22:01:44     Dump the side-table for tag: 0 with group count: 28 into file: file:/tmp/hadoop/f176cd16-bdb5-42de-a62b-5c715e541a1d/hive_2021-05-30_2
2-01-26_741_8906504297828992790-1/-local-10004/HashTable-Stage-4/MapJoin-mapfile00--.hashtable
2021-05-30 22:01:44     Uploaded 1 File to: file:/tmp/hadoop/f176cd16-bdb5-42de-a62b-5c715e541a1d/hive_2021-05-30_22-01-26_741_8906504297828992790-1/-local-10
004/HashTable-Stage-4/MapJoin-mapfile00--.hashtable (1333 bytes)
2021-05-30 22:01:44     End of local task; Time Taken: 0.808 sec.
Execution completed successfully
MapredLocal task succeeded
Launching Job 2 out of 2
Number of reduce tasks is set to 0 since there's no reduce operator
Starting Job = job_1622436790415_0003, Tracking URL = http://Master:8088/proxy/application_1622436790415_0003/
Kill Command = /usr/local/hadoop/bin/hadoop job  -kill job_1622436790415_0003
Hadoop job information for Stage-4: number of mappers: 1; number of reducers: 0
2021-05-30 22:01:50,209 Stage-4 map = 0%,  reduce = 0%
2021-05-30 22:01:57,389 Stage-4 map = 100%,  reduce = 0%, Cumulative CPU 1.97 sec
MapReduce Total cumulative CPU time: 1 seconds 970 msec
Ended Job = job_1622436790415_0003
MapReduce Jobs Launched:
Stage-Stage-1: Map: 1  Reduce: 1   Cumulative CPU: 3.69 sec   HDFS Read: 9695 HDFS Write: 246 SUCCESS
Stage-Stage-4: Map: 1   Cumulative CPU: 1.97 sec   HDFS Read: 5916 HDFS Write: 72 SUCCESS
Total MapReduce CPU Time Spent: 5 seconds 660 msec
OK
20150109        99
20160103        37
20170109        29
20180105        29
20190107        32
20200103        17
Time taken: 31.698 seconds, Fetched: 6 row(s)
```

图 6-36　一年中出现最高温度的那一天

分析：在求出每年最高温度的基础上，把它作为一张虚拟表，连接原表，通过已经求出的最高温度使其等于原表的温度，从而求出最高温度的对应时间，就求出了一年中出现最高温度的那一天。

```
select substring(b.tem, 1, 8) as max_temp_date, a.max_temp
    from tem b
join
(select substring(c.tem, 1, 4) as year,max(substring(c.tem, -2)) as max_
temp from tem c group by
    substring(c.tem, 1, 4)) a
on a.year = substring(b.tem, 1, 4)
and a.max_temp = substring(b.tem, -2);
```

任务 2　使用 Hadoop 及其组件 Spark 实现数据分析

本任务使用 spark-shell 实现词频统计。

1）数据准备，如图 6-37 所示。

#创建目录和文件

```
Hadoop@Master:/usr/local/Hadoop$ mkdir -p /usr/local/spark/data/sparkdatas
Hadoop@Master:/usr/local/Hadoop$ cd /usr/local/spark/data/sparkdatas/
Hadoop@Master:/usr/local/spark/data/sparkdatas$ vim wordcount.txt
Hadoop@Master:/usr/local/spark/data/sparkdatas$ cat wordcount.txt
```

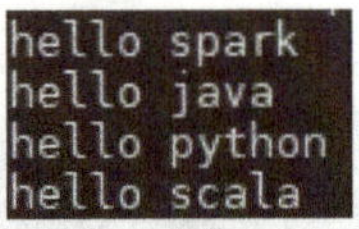

图 6-37　数据准备

2）运行 Hadoop 的 HDFS 程序，如图 6-38 所示。

　　Hadoop@Master:/usr/local/Hadoop$ sbin/start-dfs.sh
　　Hadoop@Master:/usr/local/Hadoop$ sbin/start-yarn.sh

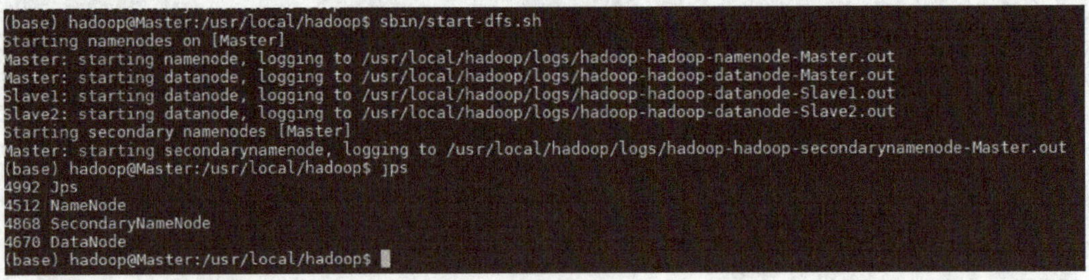

图 6-38　运行 Hadoop 的 HDFS 程序

3）运行 Hadoop 的 YARN 程序，如图 6-39 所示。

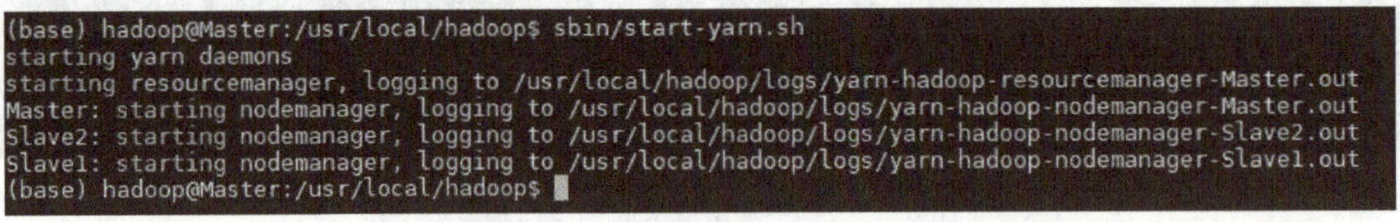

图 6-39　运行 Hadoop 的 YARN 程序

4）运行 spark-shell 程序的--master 启动本地模式，如图 6-40 所示。

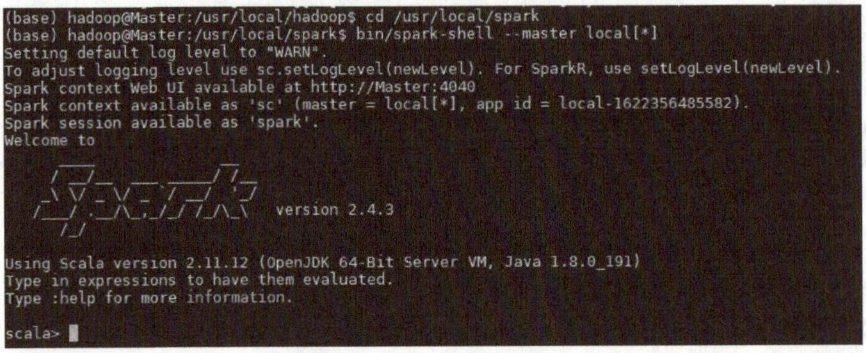

图 6-40　运行 spark-shell 程序

5）编写 Scala 脚本，实现 wordcount.txt 的词频统计。

　　scala>sc.textFile("file:///usr/local/spark/data/sparkdatas/wordcount.txt
").flatMap(x => x.split(" ")).
　　　　map(x => (x,1)).reduceByKey((x,y) => x + y).collect
　　res0: Array[(String, Int)] = Array((scala,1), (python,1), (hello,4), (java,1), (spark,1))

练习题

(1) 简述 Hadoop 框架。
(2) 简述 Hadoop 核心组件和工作原理。
(3) 简述 Hive、Spark 和 HBase 的特点。
(4) 简述 Kafka、Flume、Sqoop 和 Zookeeper 的组织架构。

项目 7　基于 scikit-learn 机器学习库的数据分析

【项目分析】

本项目旨在理解和掌握基于 scikit-learn（简称 sklearn）机器学习库的数据分析工具的基本概念和基础语法。具体内容如下。

1）机器学习简介、实施步骤、开发流程。
2）机器学习常用算法。
3）scikit-learn 机器学习库的数据分析案例介绍。

【知识准备】

7.1　掌握机器学习基本概念

7.1.1　机器学习简介

机器学习（Machine Learning，ML）是一门综合了多种理论的交叉学科，包括统计学、概率论和算法分析等。机器学习是数据分析领域的一个热点。基于大数据的机器学习则更加流行，因为其通过对数据的计算，可以实现数据预测、为公司提供决策依据。机器学习能够赋予计算机学习的能力。从逻辑上看，机器学习的应用方式是通过训练业务数据模型实现业务结果预测，这与人类的归纳经验和总结规律是完全一致的。机器学习是通过现有数据归纳出模型训练之后的一般规律，然后预测未来类似数据的发展结果。从实践的意义上来说，机器学习是一种通过利用数据，导入指定模型，不断优化模型，然后使用模型预测的一种方法，如图 7-1 所示。

图 7-1　机器学习和人类学习的过程图

机器学习的发展大致分为了三个时期：推理期、知识期和机器学习时期。每个时期都有各自的特点，如表 7-1 所示。

表 7-1　机器学习的发展

时期	特点
推理期（1960—1969）	通过计算机实现逻辑推理能力，并证明了一些数学定理
知识期（1970—1979）	通过人类的知识经验总结转换成数据传送给计算机实现专家系统
机器学习时期（1980 年至今）	通过连接主义的感知机和神经网络、深度神经网络、统计学习的支持向量机实现机器学习，特别是深度神经网络当前应用广泛

7.1.2　机器学习基本流程

机器学习的基本流程是通过建立相应的原始模型，导入指定的业务数据，不断优化模型与期望目标值的差距从而找到符合真实业务需求的规律，最终实现对未来数据及其规律的精准预测，如图 7-2 所示。

图 7-2　机器学习基本流程图

机器学习实施具体步骤如下，如图 7-3 所示。

1）针对具体的业务需求，选择合适的基本模型。模型就是一组用于处理具体业务数据的函数。

2）模型在数据的训练下不断被优化，衡量这组函数质量优劣的标准就叫作损失函数。损失函数适用于不同的具体业务需求，回归问题可以使用平方误差损失函数和绝对值误差损失函数，分类问题可以使用交叉熵损失函数。

3）在具体业务场景中找到一个最合适的模型是机器学习的关键，常用的方法有梯度下降法、最小二乘法实现局部或全局最优。

4）将最合适的模型应用到该业务的最新数据中，检验其效果。

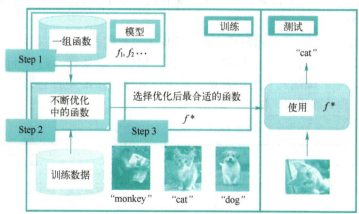

图 7-3　机器学习实施具体步骤图

7.1.3 机器学习开发流程

机器学习开发流程包括以下几步,如图7-4所示。

1)获取数据:从指定数据源获取业务数据。

2)数据处理:对源数据进行数据清理(处理缺失值、重复值和逻辑错误值)、数据集成(处理实体数据识别和冗余问题)、数据规约(尽可能保护数据原貌的前提下,最大限度地精简数据量)。

3)特征工程:数据和特征决定了机器学习的上限,而模型和算法只是逼近这个上限而已,其目的是最大限度地从原始数据中提取特征以供算法和模型使用。

4)机器学习算法训练——模型:通过导入指定的特征数据,对模型进行训练和优化。

5)模型评估:使用损失函数对模型进行评估。

6)应用:将测试数据导入评估后的模型中进行检验。

图 7-4　机器学习开发流程图

7.1.4 机器学习算法分类

机器学习算法主要分为两大类:有监督学习和无监督学习。有监督学习带有目标数据(即有标准答案),无监督学习不需要目标数据(无标准答案)。有监督学习可以再分为回归和分类,无监督学习可以再分为降维和聚类。

在有监督学习中,回归用于预测线性连续数据,主要代表算法是线性回归;分类用于预测离散数据,主要代表算法包括:支持向量机、决策树、逻辑斯谛回归、朴素贝叶斯、最近邻居。在无监督学习中,降维的主要代表算法是主成分分析;聚类的主要代表算法有:层级聚类、K均值聚类法和关联规则,如表7-2所示。

表 7-2　监督学习和无监督学习

学习方式	处理方式	主要代表算法名称
监督学习	回归	线性回归
	分类	支持向量机、决策树、逻辑斯谛回归、朴素贝叶斯、最近邻居
无监督学习	降维	主成分分析
	聚类	层级聚类、K均值聚类和关联规则

下面介绍关于机器学习的常用算法。

1. 线性回归

（1）基本思想

回归分析是一种预测性的建模技术，线性回归预测是基于某个变量 X（自变量）来预测变量 Y（因变量）的值，当然前提是 X 和 Y 之间存在线性关系。这两个变量之间的线性关系可以用直线表示（称为回归线）。这种函数是一个或多个称为回归系数的模型参数的线性组合。只有一个自变量的情况称为一元回归，大于一个自变量的情况叫作多元回归。这种技术通常用于预测分析、时间序列模型，以及发现变量之间的因果关系。通常使用曲线/直线来拟合数据点，目标是使线到数据点的距离差异最小，如图 7-5 所示是一个简单的一元线性回归效果图。

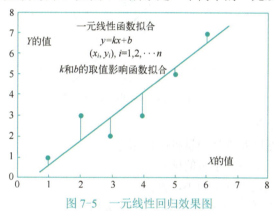

图 7-5 一元线性回归效果图

（2）线性回归特点
- 思想简单，容易实现，特别适合数据量不大，关系简单的数据集合。
- 容易理解，可读性强，特别适合决策分析。
- 可以作为许多非线性模型的基础使用。
- 能够很好地处理回归问题，如股票预测。
- 不适合处理非线性数据或者数据特征具有相关性的多项式。
- 不适合处理表达高维度的复杂数据。

（3）线性回归模型

$$h(x) = w_1x_1 + w_2x_2 + w_3x_3 + \cdots + w_nx_n + b \tag{7-1}$$

（4）线性回归注意事项

从前面的线性回归模型公式可以看出，该模型可以有多个变量。因此，线性回归得出的模型不一定是一条直线。
- 在只有一个变量时，模型是平面中的一条直线。
- 有两个变量时，模型是空间中的一个平面。
- 有更多变量时，模型将是更高维的。

2. 支持向量机

（1）基本思想

支持向量机（Support Vector Machine，SVM）是有监督学习算法的一种，用于解决数据挖掘或模式识别领域中的数据分类问题。

SVM 算法即寻找一个分类器使得超平面和最近的数据点之间的分类边缘（超平面和最近的数据点之间的间隔被称为分类边缘）最大。SVM 算法通常认为分类边缘越大，平面越优。通常定义具有"最大间隔"的决策面就是 SVM 要寻找的最优解，如图 7-6 所示。

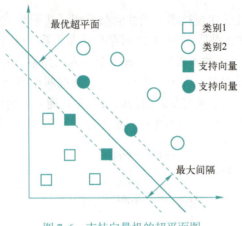

图 7-6　支持向量机的超平面图

（2）支持向量机的特点
- 最优化区分特征空间是支持向量机的目标。
- 最大化边界分类是支持向量机的核心。
- 支持向量机模型利用内存空间小，鲁棒性强。
- 支持向量机模型不适合处理数据体量较大的数据。
- 支持向量机模型不适合处理多分类问题。

（3）SVM 模型

SVM 模型可以分为三类：完全线性分割、非完全线性分割和非线性分割。

- 完全线性分割：使用 SVM 模型将不同样本完全分割，得到一个线性的结果，如图 7-7 所示。完全线性分割线的函数公式为 $w^T x+b=0$。式中，w 表示向量；T 表示转置；b 表示实数。

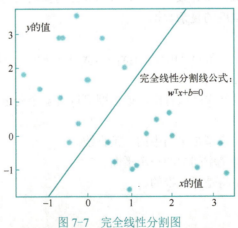

图 7-7　完全线性分割图

- 非完全线性分割：使用 SVM 模型将不同样本大部分分割，得到一个线性的结果，如图 7-8 所示。非完全线性分割线的公式为 $w^T x+b=1$，$w^T x+b=-1$ 或 $w^T x+b=0$。其中，$w^T x+b=1$ 和 $w^T x+b=-1$ 分别表示距离完全线性分割线（$w^T x+b=0$）最近的点为支持向量

（图 7-8 中虚线所穿过的点），同时定义正负值（1 和-1）表示支持向量的距离为最大边界值（Max Margin）。

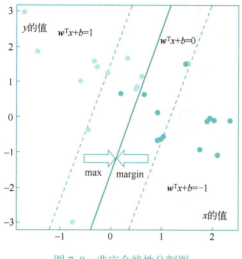

图 7-8　非完全线性分割图

3．决策树

（1）基本思想

决策树呈树形结构，在分类问题中，表示基于特征对实例进行分类的过程。决策树可以分为两类：离散性决策树和连续性决策树。离散性决策树，其目标变量是离散的，如性别。连续性决策树，其目标变量是连续的，如年龄、价格等。

（2）决策树组成元素

决策树的几个组成元素如图 7-9 所示。

- 根结点：表示整个数据样本的原点，之后所有的结点都是从这个原点出发的。
- 分离：表示将上一级结点分离成多个子结点的动作。
- 决策结点：如果某结点被继续分为多个子结点，则该结点被称为决策结点。
- 叶结点：如果某结点没有被继续分为多个子结点，则该结点被称为叶结点。
- 结点修剪：删除决策树中子结点的过程叫作结点修剪，与分离相反。
- 子树或分支：决策树中的一部分结点集合就是一个子树或分支。
- 父结点和子结点：一个结点被拆分成多个子结点则称为父结点。

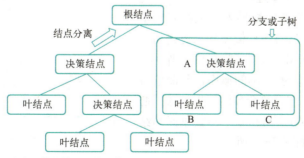

图 7-9　决策树组成结构图

(3) 决策树的特点
- 可读性和可理解性较好。
- 分类效率高，适合处理数据规模较大的高维数据。
- 容易出现过拟合现象，泛化能力较差。
- 分类规则准确性高。
- 可以处理连续和多种类字段。

(4) 决策树计算过程

决策树的计算过程示例如图 7-10 所示。
1）观察西瓜颜色是否绿色作为起始条件。
2）观察西瓜形状是否圆润。
3）西瓜的敲击声是否闷响。
4）最后得出分类结果：好西瓜。

图 7-10　决策树的基本流程

4．逻辑斯谛回归

(1) 基本思想

逻辑斯谛回归是一种分类算法，用于处理因变量为分类变量的回归问题，比较适合处理二分类问题或处理多分类问题，目的是为了预测和寻找因变量的影响因素。通过历史数据的表现对未来结果发生的概率进行预测。前面的线性回归模型比较适合处理因变量是连续的变量（例如，得到的结果是数值区间型[0-1]数据），但不能很好地处理因变量是定性的变量（定性判断为 0 或者 1），逻辑斯谛回归就比较适合。例如，当需要判断一个病人是否患有糖尿病的二分类问题时，需要从年龄、性别、血压、血糖、心跳等各种指标进行综合判断。这种场景就不能使用线性回归模型，而需要使用逻辑斯谛回归模型进行定性的判断，并且该逻辑斯谛回归模型只能输出两种结果的一种（是糖尿病或不是糖尿病）。

(2) 逻辑斯谛回归的特点
- 计算速度快，计算量与特征数量相关。
- 特征权重清晰，可读性和可理解性好。
- 适合二分类问题，并且不要降维特征数。
- 只关心特征值，内存资源利用率高和占用率小。
- 不适合处理非线性问题。
- 很难拟合真实业务数据分布，准确率较低。

(3) 逻辑斯谛回归计算过程

逻辑斯谛回归首先将数据导入线性回归模型中得到一个初值 z，然后将该初值 z 导入逻辑斯谛回归模型（又称 Sigmoid 函数），将初值 z 作为 x 轴的变量，返回结果作为 y 轴的概率，预测值对应的 y 值越接近 1 说明越符合预测结果，最后通过设定的阈值定性地判断 Sigmoid 函数返回的结果属于哪一类结果，如图 7-11 所示。

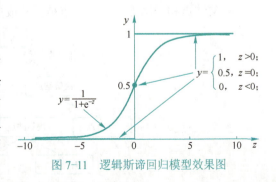

图 7-11　逻辑斯谛回归模型效果图

5. 朴素贝叶斯

（1）基本思想

朴素贝叶斯是一个常见的分类算法，比较适用于垃圾邮件过滤、文本分类或拼写查错等领域。朴素贝叶斯的最大特点就是：特征条件独立假设。该假设独立看待当前每一个条件指标与待分类项的概率关系，并不关心各个条件指标之间有无依赖关系。例如，一般情况下，银行会根据客户的月收入、信用卡额度、房车情况等不同方面的特征综合判断是否给客户办理贷款手续。因为一般对于月收入较高的客户来说，信用卡额度和房车情况都是比较良好的状态。月收入和信用卡额度、房车情况存在很高的相关性。然而朴素贝叶斯会忽略这种特征之间的内在关系，直接认为客户的月收入与信用卡额度、房产情况之间没有任何关系，三者是各自独立的特征，而只从这三个指标各自对是否办理贷款的影响概率大小去判断。

朴素贝叶斯计算事件发生的概率是根据一个已发生事件的概率，计算另一个事件的发生概率。

（2）朴素贝叶斯的特点

- 分类的效率比较稳定、可靠。
- 适合处理小规模数据和多分类任务。
- 适合增量式操作，可应对数据超出内存容量的场景。
- 对数据不太敏感，适合文本分类、垃圾邮件和拼写纠错等。
- 由于特征条件独立假设的原因，往往有悖于真实的特征之间存在较强关联的客观事实，因此比较适用于特征之间关联较小的场景。

（3）朴素贝叶斯模型

朴素贝叶斯模型，见式（7-2）。

$$P(B|A) = \frac{P(B)P(B|A)}{P(A)} \tag{7-2}$$

- $P(B)$表示发生 B 事件的概率。
- $P(A)$表示发生 A 事件的概率。
- $P(B|A)$表示在 A 事件已经发生的情况下 B 事件会发生的概率。
- $P(A|B)$表示在 B 事件已经发生的情况下 A 事件会发生的概率。

6. 最近邻居

（1）基本思想

最近邻居是一种分类和回归的非参数方法。最近邻居选择一个样本在有限范围内与邻居样本距离最近的 K 个点，通过判断这 K 个点属于什么类型，然后根据少数服从多数的原则，决定该样本属于数量较多的那一类。因此，在开展最近邻居之前，必须首先区分好这些邻居样本的分类，然后再划定目标样本和邻居样本的距离范围大小。因此，该方法在分类过程中，起到决定因素的就是相邻范围内的局部样本类型。

（2）最近邻居的特点

- 模型简单，可理解，好实现。
- 容错率较好，对噪声数据不敏感。
- 适合具有多指标对象的多分类问题。
- 根据划定 k 值的 d 数量大小，计算与每一个邻居样本距离的工作量较大。
- 不适合数据样本分布极度不平衡的场景，即一个样本数量过多或者局部样本数量过少都有可能导致判断失败。

（3）最近邻居计算过程

最近邻居的计算过程示例如图 7-12 所示。以下样本数据中有两类已经明确的数据类型：正方形和三角形。中间的圆形是待分类的目标数据。

1）以绿色圆形待分类的目标数为圆心根据距离自定义设定周围样本数量 $k=3$ 或 5 等。

2）分别根据 k 值所划定的范围，确定数据类型最多的一类。

3）当 $k=3$ 时，有两个红色三角形，一个蓝色正方形，因此绿色圆形被分为红色三角形。

图 7-12　最近邻居的计算过程

4）当 $k=5$ 时，有两个三角形，三个正方形，因此圆形被归为正方形。

7. 主成分分析

（1）基本思想

主成分分析通过找到原始数据集中最符合当前业务需求的核心数据指标，再重新组合之后得到新的指标集合，在不过多影响原始数据对业务需求影响的前提下，达到降维和减少计算量的目的。主成分分析在缩减数据、减少冗余和消除数据噪声等方面应用广泛。其中包含以下几个基本术语。

- 方差：度量一组数据分散的程度，是各个样本与样本均值的差的平方和的均值。
- 协方差：度量两个变量变动的同步程度，也就是度量两个变量线性相关性程度。如果两个变量的协方差为 0，则二者线性无关。如果大于 0 表示正相关，小于 0 表示负相关。当协方差大于 0 时，一个变量增大时另一个变量也会增大。当协方差小于 0 时，一个变量增大时另一个变量会减小。
- 协方差矩阵：由数据集中两两变量的协方差组成。矩阵的第 (x,y) 个元素是数据集中第 x 和第 y 个元素的协方差。
- 中心化：是指变量减去它的均值。

（2）主成分分析的特点

- 以方差大小作为判断影响程度，不受数据集以外的因素影响。
- 各主成分之间正交，可消除原始数据成分间的相互影响的因素。
- 主要运算是特征值分解，易于实现。
- 综合后的主成分没有原始数据完整，存在一定内涵的损失。
- 方差小的成分也有可能存在重要信息。

（3）主成分分析计算过程

主成分分析计算过程如图 7-13 所示。

1）假设有 m 个 n 维的样本数据库。
2）将原始数据按列组成 n 行 m 列矩阵 X。
3）对矩阵 X 的每行中心化。
4）求出矩阵 X 的协方差矩阵。
5）求出协方差矩阵的特征值。
6）根据特征值大小从上到下按行排列。
7）计算排序后特征值的贡献率大小。

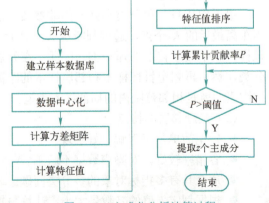

图 7-13　主成分分析计算过程

8）设定阈值决定提取主成分的个数。

8. 层级聚类

（1）基本思想

层级聚类是把数据划分成不同层级的数据结合，具有层次性。根据分层方式可以将层级聚类分为两种：分裂法和凝聚法。

- 分裂法：将所有数据样本看成一个整体，通过设定分裂规则，实现自顶向下的层层分裂，直到达到终止条件。
- 凝聚法：将每一个数据样本视为一个独立个体，通过设定合并规则，实现自底向顶的逐渐融合，直到达到终止条件。

（2）层级聚类的特点

- 分裂或凝聚的规则自由灵活，容易实现。
- 层级关系明显、清晰、可理解。
- 层级数量不受限制。
- 由于规则灵活，所以计算复杂度高。
- 异常值对结果影响很大。

（3）层级聚类计算过程

层级聚类的计算过程如图 7-14 所示。

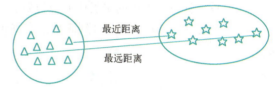

图 7-14　层级聚类算法计算过程

1）分裂法。

- 所有数据样本视为一个集合。
- 设定集合中数据样本之间的距离阈值作为分裂标准。
- 在同一个集合中计算样本之间的距离，找出距离最远的两个样本 X 和 Y，并将其分为两类。
- 计算集合中剩余样本分别与样本 X 和 Y 的距离。如果离 X 近，则为 X 同类。反之，则为 Y 同类。
- 达到设定的聚类条件为止。

2）凝聚法。

- 集合中每个样本都视为一个类别。
- 设定集合中数据样本之间的距离最小的两个样本距离作为凝聚标准。
- 在同一个集合中计算样本之间的距离，找出距离最小的两个样本 X 和 Y，并将其凝聚为一类。
- 同理，计算集合剩余样本中距离最小的两个样本，并凝聚为一类。
- 达到设定的聚类条件为止。

9. K 均值聚类法

（1）基本思想

在数据样本中首先设定 K 个样本点作为聚类的中心点，然后通过对比每个中心点与周围数据点距离的大小，决定数据点归属的类别，把每个数据点分配给距离它最近的聚类中心。根据这种聚类方式，每次吸纳了新的数据点后，便重新计算当前聚类中心的位置（各数据点到中心点的平均距离），直到满足指定的聚类终止条件为止。终止条件包括：没有（或最小数目）数据点被分配给不同的聚类、聚类中心不再发生变化或达到误差平方和局部最小。

(2) K 均值聚类法的特点
- 算法收敛速度快，容易实现。
- 可理解性和可读性好。
- 参数简单，只需设置聚类中心 K 的数量。
- 聚类中心 K 的数量需要认真思考。
- 不适合处理噪声较多的数据。
- 不适合处理类别数量严重不平衡的数据集合。

(3) K 均值聚类计算过程

K 均值聚类的计算过程如图 7-15 所示。

1) 根据具体业务数据选择 K 个数据点作为聚类中心点。
2) 计算所有数据点与 K 个数据点的距离。
3) 将离聚类中心点最近的数据点归为同一类。
4) 重新计算新类中的聚类中心点位置。
5) 重复 2)、3) 步骤，直到满足聚类终止条件。

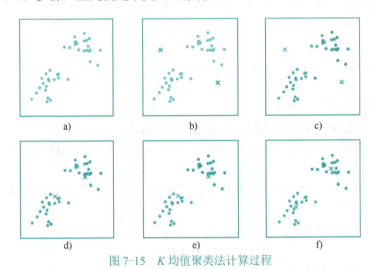

图 7-15　K 均值聚类法计算过程

a) 选择 K 个数据点作为聚类中心点　b) 计算所有数据点与 K 个数据点的距离
c) 将离聚类中心点最近的数据点归为同一类　d) 重新计算聚类中心点位置　e) 重复 b)　f) 重复 c)

10. 关联规则

(1) 基本思想

关联规则是从数据集中发现不同数据项之间可能存在的相互关联，特别适合从大规模业务数据中挖掘出有价值的业务信息。具有代表性的算法是 Apriori。关联规则比较著名的故事是"啤酒与尿布"，这是美国沃尔玛连锁店超市的真实案例：沃尔玛拥有世界上最大的数据仓库系统，集中了其各门店的详细原始交易数据。通过对消费者购物行为数据进行分析发现，男性顾客在购买婴儿尿布时，会顺便购买几瓶啤酒，于是推出了将啤酒和尿布放在一起的促销手段。揭示了一个隐藏在"尿布与啤酒"背后的美国人的一种行为模式。"啤酒＋尿布"的数据分析成果成了大数据技术与应用的经典案例，如图 7-16 所示。

图 7-16 啤酒与尿布的故事

（2）关联规则的基本术语
- 事务：一条交易信息称为一个事务。
- 项：一个交易物品称为一个项。
- 项集：包含零个或多个项的组合称为一个项集。
- K-项集：包含 K 个项的组合称为 K-项集。
- 频繁项集：设定支持度大于或等于指定阈值的项集称为频繁项集。
- 强关联规则：大于或等于最小支持度阈值和最小置信度阈值的规则叫作强关联规则。
- A 和 B 的支持度（Support）：表示同时购买 A 和 B 物品的订单数 $\text{Freq}(A \cap B)$ 占总订单数 N 的比例，即 A 和 B 物品同时被购买的概率，见式（7-3）。

$$\text{Support}(A \cap B) = \frac{\text{Freq}(A \cap B)}{N} \tag{7-3}$$

- A 对 B 的置信度（Confidence）：表示同时购买 A 和 B 物品的订单数 $\text{Freq}(A \cap B)$ 占购买 A 物品订单数 $\text{Freq}(A)$ 的比例，即购买 A 物品的顾客有多大概率会去购买 B 物品，见式（7-4）。

$$\text{Confidence} = \frac{\text{Freq}(A \cap B)}{\text{Freq}(A)} \tag{7-4}$$

- A 对 B 的提升度（Lift）：表示先购买 A 对购买 B 的概率的提升作用，在购物车中出现的次数是否高于商品单独出现在购物车中的频率，见式（7-5）。

$$\text{Lift} = \frac{\text{Support}(A \cap B)}{\text{support}(A) \times \text{support}(B)} \tag{7-5}$$

（3）Apriori 关联规则的特点
- 通过首先筛选候选集，能够很好地减少频繁集的筛选工作量。
- 每次迭代筛选过程都要遍历所有数据项，效率较低。

（4）Apriori 关联规则计算过程

Apriori 关联规则的计算过程如图 7-17 所示。

1）最小支持度设置为 50%。
2）从数据库中扫描数据。
3）生成候选频繁 1 项集 $C1$。
4）使用最小支持度阈值筛选之后最终的频繁 1 项集为 1235，即 $L1$。
5）连接生成候选频繁 2 项集，包括 12、13、15、23、25、35 共 6 组，即 $C2$。
6）计算候选频繁 2 项集 $C2$ 的支持度之后得到 $L2$。

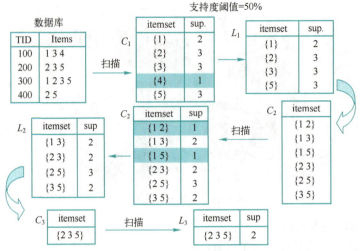

图 7-17　Apriori 关联规则计算过程

7）连接生成候选频繁 3 项集，123、125、135 和 235 共 4 组，通过支持度筛选后得到的频繁 3 项集为 235 一组，即 C3。

8）最终得到频繁 3 项集 235，即 L3。

7.2　掌握 scikit-learn 的基本用法

7.2.1　scikit-learn 的安装和引用方法

scikit-learn 是建立在 NumPy、SciPy、Pandas 和 Matplotlib 之上基于 Python 语言的机器学习工具。scikit-learn 封装了六大模块：分类、回归、聚类、降维、模型选择和预处理，如图 7-18 所示。

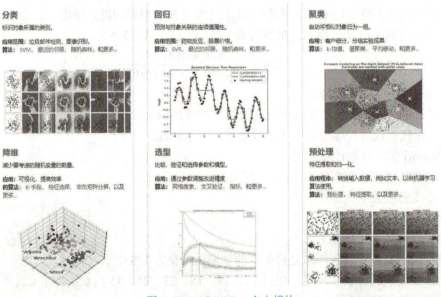

图 7-18　scikit-learn 六大模块

与前面 NumPy、Pandas 和 Matplotlib 的引用方式类似，这里通过引用 Anaconda 自带的 Python 解释器就可以导入 pandas 包，如图 7-19 所示。

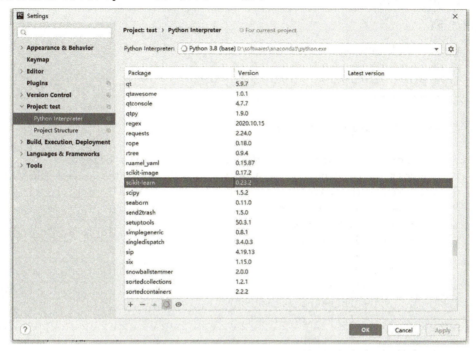

图 7-19　scikit-learn 引用

7.2.2　scikit-learn 的基本用法

scikit-learn 自带丰富的数据集，可以用来对算法进行测试分析，解决了用户寻找数据的问题，如表 7-3 所示。

表 7-3　scikit-learn 数据集

中文名称	数据集加载方法
鸢尾花数据集	load_iris()
手写数字数据集	load_digitals()
糖尿病数据集	load_diabetes()
乳腺癌数据集	load_breast_cancer()
波士顿房价数据集	load_boston()
体能训练数据集	load_linnerud()

【实例 7-1】 以鸢尾花数据集为例，导入并查看该数据基本描述信息，该数据集一共包含 4 个特征变量，1 个类别变量。共有 150 个样本，iris 是鸢尾花，这里存储了其萼片和花瓣的长宽，共 4 个属性，鸢尾花分三类。

```
# -*- coding: utf-8 -*-
#导入sklearn的数据集datasets
import sklearn.datasets as datasets
# 使用load_iris函数导入鸢尾花数据集
```

```
iris = datasets.load_iris()
# 使用 DESCR 查看鸢尾花数据集基本描述
print(iris.DESCR)
```

输出如下：

```
.. _iris_dataset:

Iris plants dataset
--------------------

**Data Set Characteristics:**

    :Number of Instances: 150 (50 in each of three classes)
    :Number of Attributes: 4 numeric, predictive attributes and the class
    :Attribute Information:
        - sepal length in cm
        - sepal width in cm
        - petal length in cm
        - petal width in cm
        - class:
                - Iris-Setosa
                - Iris-Versicolour
                - Iris-Virginica

    :Summary Statistics:

    ============== ==== ==== ======= ===== ====================
                    Min  Max   Mean    SD   Class Correlation
    ============== ==== ==== ======= ===== ====================
    sepal length:   4.3  7.9   5.84   0.83    0.7826
    sepal width:    2.0  4.4   3.05   0.43   -0.4194
    petal length:   1.0  6.9   3.76   1.76    0.9490  (high!)
    petal width:    0.1  2.5   1.20   0.76    0.9565  (high!)
    ============== ==== ==== ======= ===== ====================

    :Missing Attribute Values: None
    :Class Distribution: 33.3% for each of 3 classes.
    :Creator: R.A. Fisher
    :Donor: Michael Marshall (MARSHALL%PLU@io.arc.nasa.gov)
    :Date: July, 1988
```

【任务实施】

任务 1 使用 scikit-learn 实现鸢尾花数据分析

应用 PyCharm，以鸢尾花数据集为例，按照导入包、获取数据、数据预处理、训练模型、模型评估、预测和分类的流程，使用 svm 算法分类器实现对鸢尾花数据模型中 4 个属性的分类，以及该数据模型精确度评估，并使用持久化方式保存和运行模型。

（1）导入相应的包

```
# -*- coding: utf-8 -*-
```

```python
# sklearn 简单实现鸢尾花实例 svm 分类器
import numpy as np
import pandas as pd
import matplotlib.pyplot as plt
# 导入 sklearn 的数据集 datasets
from sklearn import datasets
```

(2) 数据的获取

```python
# 使用 load_iris 方法从数据集 datasets 导入鸢尾花数据，iris 数据包含 4 个特征变量
iris = datasets.load_iris()
# 特征变量
iris_X = iris.data
print('(1) 特征变量')
print(iris_X)
# 目标值
iris_y = iris.target
print('(2) 目标值')
print(iris_y)
# 使用 DataFrame 将鸢尾花特征变量转换为 DataFrame 对象 df，列名 columns 为鸢尾花数据 iris.feature_names
df=pd.DataFrame(iris.data,columns=iris.feature_names)
print('(3) 使用 DataFrame 将鸢尾花特征变量转换为 DataFrame 对象 df，列名 columns 为鸢尾花数据 iris.feature_names')
print(df)
# 获取 iris.target 目标值，并为 df 对象增加一个列 target
df['target']=iris.target
```

(3) 数据预处理

```python
# 导入 sklearn 的预处理模块 preprocessing
from sklearn import preprocessing
# MinMaxScaler 将样本特征值线性缩放到 0～1 之间，并返回预处理的封装对象 scaler
scaler=preprocessing.MinMaxScaler()
# 使用 fit 函数设置预处理的具体操作
scaler.fit(iris.data)
# 使用 transform 函数具体执行该预处理操作，即将 iris.data 的数据线性缩放到 0～1 之间
data=scaler.transform(iris.data)
target=iris.target
```

(4) 模型的训练

```python
# 引入 sklearn 的模型选择的库 model_selection
# 引入 model_selection 的测试集和训练集分类方法 train_test_split
from sklearn.model_selection import train_test_split
# 使用 train_test_split 方法将特征数据和目标数据按照 1：2 的比例分割成训练集和测试集
# X_train 表示特征数据的训练集，X_test 表示特征数据的测试集
# y_train 表示目标数据的训练集，y_test 表示目标数据的测试集
X_train,X_test,y_train,y_test=train_test_split(data,target,test_size=1/3)
# 导入 sklearn 的 svm 分类器
from sklearn import svm
# 使用 svm 的 SVC 类，该类主要用于数据样本较少的二元或多元分类场景
# kernel='linear' 表示线性核，是最简单的核函数
# C 为惩罚参数，C 越大，对误分类的惩罚越大，反之，惩罚越小
```

```
clf = svm.SVC(kernel='linear',C=1)
# 用 fit 函数设置使用训练集数据的训练模型
clf.fit(X_train,y_train)
# 用 predict 函数执行特征数据的测试集，并返回预测结果
a=clf.predict(X_test)
# 查看模型的参数
b=clf.get_params()
# 查看匹配度，即使用 svm 分类模型训练后，训练集和测试集结果的正确率
c=clf.score(X_test,y_test)
print('（4）查看匹配度，即使用 svm 分类模型训练后，训练集和测试集结果的正确率')
print(c)
```

（5）模型的评估

```
# 导入 sklearn.metrics 中的 classification_report 分类模型评分报告用于评估模型质量
from sklearn.metrics import classification_report
# 使用 classification_report 函数将 target 目标值和模型的预测值 clf.predict(data)进行分析报告
print('（5）使用 classification_report 函数将 target 目标值和模型的预测值 clf.predict(data)进行分析报告')
print(classification_report(target,clf.predict(data),target_names=iris.target_names))
# 导入 sklearn.model_selection 的 cross_val_score 交叉验证函数，用于更加准确地评估模型质量
from sklearn.model_selection import cross_val_score
# 使用 cross_val_score 函数将 svm 分类器对象 clf，预处理后的数据 data 以及目标值 target，按照 cv=5，即分成 5 份进行交叉验证
scores= cross_val_score(clf,data,target,cv=5)
print('（6）输出交叉验证后的模型评估得分')
print(scores)
```

（6）模型持久化

```
# 导入 pickle 用于保存数据
import pickle
# 使用 dumps 函数保存 svm 模型成为字符串
s = pickle.dumps(clf)
# 使用 loads 函数从字符串加载模型
clf2=pickle.loads(s)
# 使用 predict 函数运行 clf2 模型的数据 data
z=clf2.predict(data)
print('（7）使用 predict 函数运行 clf2 模型的数据 data')
print(z)
```

（7）输出结果

（1）特征变量

```
[[5.1 3.5 1.4 0.2]
 [4.9 3.  1.4 0.2]
 [4.7 3.2 1.3 0.2]
 [4.6 3.1 1.5 0.2]
 [5.  3.6 1.4 0.2]
 [5.4 3.9 1.7 0.4]
 [4.6 3.4 1.4 0.3]
```

```
 [5.  3.4 1.5 0.2]
 [4.4 2.9 1.4 0.2]
 [4.9 3.1 1.5 0.1]]
```

(2) 目标值

```
[0 0 0 0 0 0 0 0 0 0 0 0 0 0 0 0 0 0 0 0 0 0 0 0 0 0 0 0 0 0 0 0 0 0 0
 0 0 0 0 0 0 0 0 0 0 0 0 0 0 0 1 1 1 1 1 1 1 1 1 1 1 1 1 1 1 1 1 1 1 1
 1 1 1 1 1 1 1 1 1 1 1 1 1 1 1 1 1 1 1 1 1 1 1 1 1 1 1 1 1 1 2 2 2 2 2 2 2 2 2 2
 2 2 2 2 2 2 2 2 2 2 2 2 2 2 2 2 2 2 2 2 2 2 2 2 2 2 2 2 2 2 2 2 2 2 2 2 2 2 2
 2 2]
```

(3) 使用 DataFrame 将鸢尾花特征变量转换为 DataFrame 对象 df，列名 columns 为鸢尾花数据 iris.feature_names

```
     sepal length (cm)  sepal width (cm)  petal length (cm)  petal width (cm)
0           5.1               3.5               1.4                0.2
1           4.9               3.0               1.4                0.2
2           4.7               3.2               1.3                0.2
3           4.6               3.1               1.5                0.2
4           5.0               3.6               1.4                0.2
..          ...               ...               ...                ...
145         6.7               3.0               5.2                2.3
146         6.3               2.5               5.0                1.9
147         6.5               3.0               5.2                2.0
148         6.2               3.4               5.4                2.3
149         5.9               3.0               5.1                1.8

[150 rows x 4 columns]
```

(4) 查看匹配度，即使用 svm 分类模型训练后，训练集和测试集结果的正确率
0.94

(5) 使用 classification_report 函数将 target 目标值和模型的预测值 clf.predict(data) 进行分析报告

```
              precision    recall  f1-score   support

      setosa       1.00      1.00      1.00        50
  versicolor       0.91      1.00      0.95        50
   virginica       1.00      0.90      0.95        50

    accuracy                           0.97       150
   macro avg       0.97      0.97      0.97       150
weighted avg       0.97      0.97      0.97       150
```

(6) 输出交叉验证后的模型评估得分

```
[0.96666667 0.96666667 0.96666667 0.93333333 1.        ]
```

(7) 使用 predict 函数运行 clf2 模型的数据 data

```
[0 0 0 0 0 0 0 0 0 0 0 0 0 0 0 0 0 0 0 0 0 0 0 0 0 0 0 0 0 0 0 0 0 0 0
 0 0 0 0 0 0 0 0 0 0 0 0 0 0 0 1 1 1 1 1 1 1 1 1 1 1 1 1 1 1 1 1 1 1 1
 1 1 1 1 1 1 1 1 1 1 1 1 1 1 1 1 1 1 1 1 1 1 1 1 1 1 1 2 2 2 2 1 2 2 2
 2 2 2 2 2 2 2 1 2 2 2 2 2 2 2 2 2 1 1 2 2 2 2 2 2 2 2 2 2 2 2 2 2 2
 2 2]
```

任务 2　使用 scikit-learn 实现波士顿房价数据分析

房价的高低是受多个因素影响的，如房子所处的城市是一线还是二线，房子周边交通方便程度，如通不通地铁、房子周边有无学校和医院等。应用 Jupyter Notebook，以波士顿房价数据集为例，按照导入包、获取数据、数据预处理、训练模型的流程，使用线性回归分析波士顿房价数据集中的不同属性与房价的相关性，并通过训练的模型预测房价，如下所示。

（1）导入相应的包

```
# -*- coding: utf-8 -*-
# 导入 mean_squared_error 的均方误差计算方法
from skimage.metrics import mean_squared_error
# 导入 sklearn 的波士顿房价数据集获取方法
from sklearn.datasets import load_boston
# 导入 sklearn 的线性回归类
from sklearn.linear_model import LinearRegression
# 导入 sklearn 的模型选择包的训练数据和测试数据分类方法
from sklearn.model_selection import train_test_split
# 导入 sklearn 的标准化类
from sklearn.preprocessing import StandardScaler
# 导入 Matplotlib 图形库
from matplotlib import pyplot as plt
# 导入 Pandas 库
import pandas as pd
# 导入 NumPy 库
import numpy as np
# 设置图形库字体格式，避免出现乱码
plt.rcParams['font.sans-serif']=['SimHei']
plt.rcParams['axes.unicode_minus']=False
```

（2）获取数据

```
# 波士顿房价数据说明：此数据源于美国某经济学杂志，分析研究波士顿房价(Boston
# HousePrice)的数据集，如图 7-20 所示。
lb = load_boston()
Lb.data
```

```
Out[11]: array([[6.3200e-03, 1.8000e+01, 2.3100e+00, ..., 1.5300e+01, 3.9690e+02,
          4.9800e+00],
         [2.7310e-02, 0.0000e+00, 7.0700e+00, ..., 1.7800e+01, 3.9690e+02,
          9.1400e+00],
         [2.7290e-02, 0.0000e+00, 7.0700e+00, ..., 1.7800e+01, 3.9283e+02,
          4.0300e+00],
         ...,
         [6.0760e-02, 0.0000e+00, 1.1930e+01, ..., 2.1000e+01, 3.9690e+02,
          5.6400e+00],
         [1.0959e-01, 0.0000e+00, 1.1930e+01, ..., 2.1000e+01, 3.9345e+02,
          6.4800e+00],
         [4.7410e-02, 0.0000e+00, 1.1930e+01, ..., 2.1000e+01, 3.9690e+02,
          7.8800e+00]])
```

图 7-20　获取波士顿房价数据

（3）分析数据特征

```
# 波士顿房价数据集包含 506 个样本和 13 个特征：(506, 13)
print(lb.data.shape)

# 数据集中的每一行数据都是对波士顿周边或城镇房价的情况描述，下面对数据集变量进行说明
# 13 个特征名称
# CRIM：城镇人均犯罪率
# ZN：住宅用地所占比例
# INDUS：城镇中非住宅用地所占比例
# CHAS：虚拟变量，用于回归分析
# NOX：环保指数
# RM：每栋住宅的房间数
# AGE：1940 年以前建成的自住单位的比例
```

```
# DIS：与 5 个波士顿的就业中心的加权距离
# RAD：距离高速公路的便利指数
# TAX：每一万美元的不动产税率
# PTRATIO：城镇中的教师、学生比例
# LSTAT：地区中有多少房东属于低收入人群
# ['CRIM' 'ZN' 'INDUS' 'CHAS' 'NOX' 'RM' 'AGE' 'DIS' 'RAD' 'TAX' 'PTRATIO' 'LSTAT']
print(lb.feature_names)
```

（4）数据可视化

通过散点图呈现，遍历所有 13 个特征与房价的相关系数，如图 7-21～图 7-32 所示。

```
for i in range(13):
    # 获取每一个特征的列值
    x = lb.data[:,np.newaxis,i]
    # 获取房价目标值
    y = lb.target
    # 初始化线性回归对象
    lr_each = LinearRegression()
    # 传入参数
    lr_each.fit(x,y)
    # 决定系数：coef_反映了 y 的波动有多少百分比能被 x 的波动所描述，即表征因变量 y 的变异中有多少百分比可由控制的自变量 x 来解释
    # 决定系数：coef_越大，说明 x 对 y 的解释程度越高
    print(lb.feature_names[i]+"特征与房价的相关系数为："+str(lr_each.coef_))
    # 使用散点图分别呈现 13 个特征与房价的相关系数
    plt.scatter(x,y,color='red')
    plt.plot(x,lr_each.predict(x),color='yellow',linewidth=6)
    plt.xlabel(lb.feature_names[i])
    plt.ylabel('房价')
    plt.title('线性回归房价格与'+lb.feature_names[i]+'的关系')
plt.show()
```

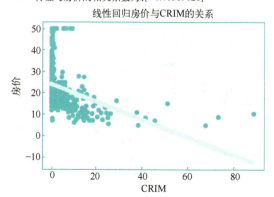

图 7-21　线性回归房价与 CRIM 的关系

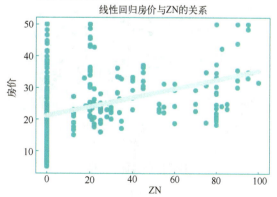

图 7-22　线性回归房价与 ZN 的关系

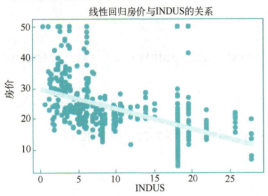

图 7-23 线性回归房价与 INDUS 的关系

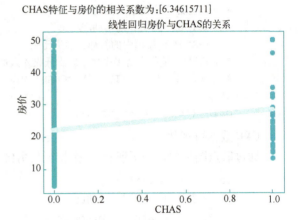

图 7-24 线性回归房价与 CHAS 的关系

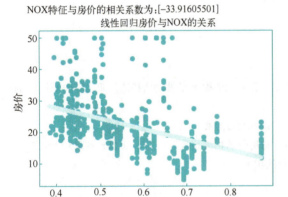

图 7-25 线性回归房价与 NOX 的关系

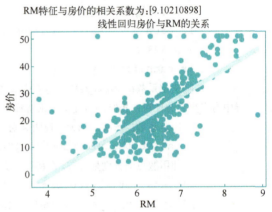

图 7-26 线性回归房价与 RM 的关系

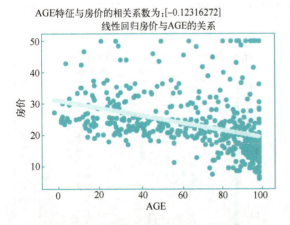

图 7-27 线性回归房价与 AGE 的关系

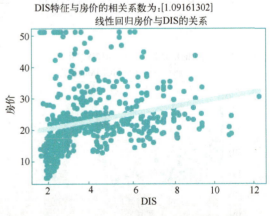

图 7-28 线性回归房价与 DIS 的关系

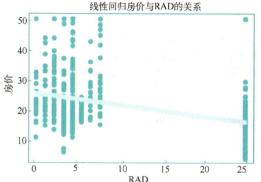

图 7-29 线性回归房价与 RAD 的关系

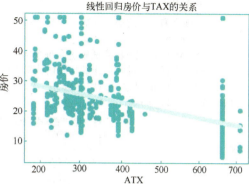

图 7-30 线性回归房价与 TAX 的关系

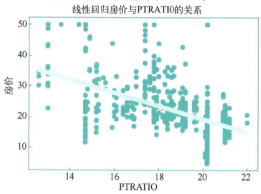

图 7-31 线性回归房价与 PTRATIO 的关系

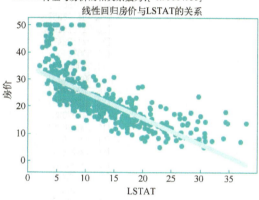

图 7-32 线性回归房价与 LSTAT 的关系

（5）数据预处理

```
    # 分割数据集到训练集和测试集
    x_train, x_test, y_train, y_test = train_test_split(lb.data, lb.target,
test_size=0.25)
    # print(y_train, y_test)

    # 特征值和目标值都必须进行标准化处理
    # 特征值标准化
    std_x = StandardScaler()
    x_train = std_x.fit_transform(x_train)
    x_test = std_x.transform(x_test)

    # 目标值标准化
    std_y = StandardScaler()
    y_train = std_y.fit_transform(y_train.reshape(-1, 1))
    y_test = std_y.transform(y_test.reshape(-1, 1))
```

（6）模型训练

```
    # 正规方程求解方式预测
    lr = LinearRegression()
    lr.fit(x_train, y_train)
```

(7) 模型预测

```
# 预测测试集的房子价格，如图 7-33 所示。
y_lr_predict = std_y.inverse_transform(lr.predict(x_test))
print("正规方程测试集里面每个房子的预测价格：", y_lr_predict)
# 均方误差是反映估计量与被估计量之间差异程度的一种度量，该值越大越好，如图 7-34 所示。
print("正规方程的均方误差：", mean_squared_error(std_y.inverse_transform(y_test), y_lr_predict))
```

```
正规方程测试集里面每个房子的预测价格： [[30.84089306]
 [17.03173868]
 [21.64661517]
 [15.41548366]
 [26.00868978]
 [22.40145932]
 [20.42164133]
 [24.48061365]
 [32.87623792]
 [-4.63038668]
 [34.33451825]
 [25.01721519]
 [27.62463947]
 [30.50673402]
 [24.75380654]
 [30.80352582]
 [ 8.25340464]
 [27.20759879]
 [14.44292938]
 [10.85667675]
 [21.40662563]
 [35.80821801]
 [24.89799731]
 [24.64325447]
 [15.90856431]
 [ 6.56574763]
 [19.44189932]
 [25.36133136]
 [18.93825378]
 [24.06962573]
 [20.81836999]
 [11.57617767]
 [16.57221473]
 [28.04816524]
 [24.96922022]
 [22.42544461]
 [27.32733717]
```

图 7-33　预测测试集的房子价格

```
正规方程的均方误差： 28.45907791449455
```

图 7-34　均方误差

(8) 模型持久化

```
# 导入 pickle 用于保存数据
import pickle
# 使用 dumps 函数保存 svm 模型成为字符串
s = pickle.dumps(lr)
# 使用 loads 函数从字符串加载模型
lr2 = pickle.loads(s)
# 使用 predict 函数运行 lr2 模型的数据 x_test
z = lr2.predict(x_test)
```

练习题

（1）简述机器学习的开发流程。
（2）简述机器学习的实施步骤。
（3）机器学习常用算法分为哪两大类？
（4）简述朴素贝叶斯的基本思想和特点。
（5）简述 K 均值聚类法的基本思想和计算过程。
（6）简述"啤酒与尿布"故事。

参 考 文 献

[1] 付雯，聂强. Spark 大数据实时分析实战[M]. 北京：北京理工大学出版社，2020.
[2] 李俊翰，付雯，等. 大数据采集与爬虫[M]. 北京：机械工业出版社，2020.
[3] 张靖，李俊翰，等. 大数据平台应用[M]. 北京：电子工业出版社，2020.